SpringerBriefs in Physics

SpringerBriefs in Physics are a series of slim high-quality publications encompassing the entire spectrum of physics. Manuscripts for SpringerBriefs in Physics will be evaluated by Springer and by members of the Editorial Board. Proposals and other communication should be sent to your Publishing Editors at Springer.

Featuring compact volumes of 50 to 125 pages (approximately 20,000–45,000 words), Briefs are shorter than a conventional book but longer than a journal article. Thus Briefs serve as timely, concise tools for students, researchers, and professionals.

Typical texts for publication might include:

- A snapshot review of the current state of a hot or emerging field
- A concise introduction to core concepts that students must understand in order to make independent contributions
- An extended research report giving more details and discussion than is possible in a conventional journal article
- A manual describing underlying principles and best practices for an experimental technique
- An essay exploring new ideas within physics, related philosophical issues, or broader topics such as science and society

Briefs allow authors to present their ideas and readers to absorb them with minimal time investment.

Briefs will be published as part of Springer's eBook collection, with millions of users worldwide. In addition, they will be available, just like other books, for individual print and electronic purchase.

Briefs are characterized by fast, global electronic dissemination, straightforward publishing agreements, easy-to-use manuscript preparation and formatting guidelines, and expedited production schedules. We aim for publication 8–12 weeks after acceptance.

More information about this series at http://www.springer.com/series/8902

Aniello Lampo · Miguel Ángel García March ·
Maciej Lewenstein

Quantum Brownian Motion Revisited

Extensions and Applications

 Springer

Aniello Lampo
Internet Interdisciplinary Institute
Universitat Oberta de Catalunya
Castelldefels, Barcelona, Spain

Miguel Ángel García March
QOT group
ICFO - Institute of Photonic Sciences
Castelldefels, Barcelona, Spain

Maciej Lewenstein
QOT group
ICFO - Institute of Photonic Sciences
Castelldefels, Barcelona, Spain

ICREA
Barcelona, Spain

ISSN 2191-5423   ISSN 2191-5431   (electronic)
SpringerBriefs in Physics
ISBN 978-3-030-16803-2   ISBN 978-3-030-16804-9   (eBook)
https://doi.org/10.1007/978-3-030-16804-9

Library of Congress Control Number: 2019936284

This Springer imprint is published by the registered company Springer Nature Switzerland AG
The registered company address is: Gewerbestrasse 11, 6330 Cham, Switzerland

Contents

Chapter 1
Introduction

Brownian motion refers to random dynamics of heavy particles suspended in a liquid or a gas, generated by their collisions with the constituents of the fluid around. This transport phenomenon is named after the botanist Robert Brown. In 1827, while he was looking through a microscope at pollen grains in water, he noted that the grains moved, but he was not able to determine the mechanisms that caused this motion. Atoms and molecules had long been theorized as the constituents of the matter, and many years later, Albert Einstein explained in precise detail how the motion observed by Brown was a result of the pollen being moved by individual water molecules. So, Brownian motion became an important model aimed to approach a large set of different contexts characterized by a non-deterministic behavior, or where dissipative processes occur, as result of the unavoidable interaction with the environment around (cf. Mazo 2002; Gardiner and Zoller 2004; Breuer and Petruccione 2007; Weiss 2008).

In his study of the Brownian motion, Albert Einstein considered both the pollen grains and the water molecules as classical objects, because a macroscopic particle in a viscous medium can be correctly described by the classical theory. We refer to such system as classical Brownian motion. However, in many situations of physical interest, we are not able to describe the dynamics of the system in a completely classical manner. For instance, as the temperature starts to achieve very low values, we expect to encounter quantum effects. To address these issues, a theory of quantum Brownian motion (QBM) was developed. The QBM describes the behavior of a particle coupled with a thermal[1] bath made up of a large number of quantum harmonic oscillators, satisfying the Bose-Einstein statistics. It represents a paradigmatic example of open quantum system, i.e. a system whose dynamics is affected by the surrounding non-controllable degrees of freedom, and it has been studied for

---

[1]More generally, the bath can non-necessarily be initially in thermal stationary state of some dynamics.

© The Author(s), under exclusive license to Springer Nature Switzerland AG 2019
A. Lampo et al., *Quantum Brownian Motion Revisited*, SpringerBriefs
in Physics, https://doi.org/10.1007/978-3-030-16804-9_1

decades (Caldeira and Leggett 1983a, b; Grabert et al. 1988; Hu et al. 1992; Breuer and Petruccione 2007; Schlosshauer 2007; Waldenfels 2014; Vega and Daniel 2017).

First of all, QBM is often adopted to analyze some problems lying at the core of the foundations of quantum mechanics. Here, one of the main unsolved questions regards the so-called *quantum to classical transition*, i.e. how the classical features, we experience every day in the macroscopic world, arise from the underlying quantum domain. For instance, the *non-observability of interference*: why is it so difficult to detect quantum interference effects on mesoscopic and macroscopic scales? Moreover, there is the famous *problem of outcomes*, also known as the measurement problem: what, within a measurement process, selects a particular outcome among the different possibilities described by the quantum probability distribution? Almost all the attempts to deal with these issues are based on the idea that a quantum system cannot be considered as isolated from the degrees of freedom around. Conversely, it continuously interacts with these degrees of freedom, and the quantum effects are suppressed, just as a consequence of such an interaction (Schlosshauer 2007; Zurek 2009). In this framework, QBM is often employed to describe this interaction concretely and in detail, in order to approach quantitatively the problems mentioned above (Schlosshauer 2007; Blume-Kohout and Zurek 2008; Tuziemski and Korbicz 2015; Galve et al. 2016).

In addition, QBM has been employed to investigate the origins of friction, i.e. the force resisting the relative motion of solid surfaces, fluid layers, and material elements sliding against each other. Friction is a macroscopic phenomenon, but its microscopic mechanisms are still only partially known and controversial. The theory of QBM has often been employed in this context, i.e. to construct a microscopic theory of the friction (Gardiner and Zoller 2004; Caldeira and Leggett 1983a, b). Analogous theories have been devised in quantum electrodynamics in order to explain the origins of radiation damping and Lamb shift (Rzazewski and Zakowicz 1971, 1976, 1980; Zakowicz and Rzazewski 1974; Wodkiewicz and Eberly 1976; Lewenstein and Rzazewski 1980).

Finally, QBM is also exploited for more practical purposes. For example, in many physical situations it constitutes the default choice for evaluating the decoherence and dissipation processes occurring in a system due to the interaction with the surrounding non-controllable degrees of freedom (Weiss 2008). This task is particularly important in all experiments aimed to build and detect macroscopic coherent superpositions. In this context, it is very important to monitor decoherence, because it destroys the interference effects, and QBM permits to analyze these processes providing analytical expressions for the time scales ruling them (Marshall 2003; Gröblacher et al. 2015). Obviously, the importance of decoherence and dissipation goes beyond this class of problems, and plays a very important role also in other fields, such as quantum biology: also in this case QBM provides a good scheme to describe the effects disturbing quantum features in many biological processes (Abbot et al. 2008).

The vast majority of the literature on QBM is devoted to microscopic models in which the coupling of the Brownian particle to the bosonic bath is linear both in bath creation and annihilation operators, and in position (or momentum) of the particle. The case when such coupling is non-linear in either the bath or the system operators has been investigated, for instance, in the old papers of Landauer (1957), who studied non-linearity in bath operators, and Dykman and Krivoglaz (1975), Hu et al. (1993), Brun (1993), and Banerjee et al. (2003). Physically, the case of a coupling that deviates from linearity in the system coordinates, corresponds to a situation where damping and diffusion are spatially inhomogeneous. Obviously, this non-linearity might have both classical and quantum consequences, and as such deserves careful analysis.

The inhomogeneity mentioned above has been recently intensively studied in the context of classical Brownian motion and other classical diffusive systems. In particular, explicit formulas were derived for *noise-induced drifts* in the small-mass (Smoluchowski 1916; Kramers 1940) and other limits (Hottovy et al. 2012a, 2014; McDaniel 2014). Noise-induced drifts have been shown to appear in a general class of diffusive systems, including systems with time delay and systems driven by colored noise. Applications include Brownian motion in diffusion gradient (Volpe et al. 2010; Brettschneider et al. 2011), noisy electrical circuits (Pesce et al. 2012) and thermophoresis (Hottovy et al. 2012b). In the first two cases the theoretical predictions have been demonstrated to be in an excellent agreement with the experiments. Diffusion in inhomogeneous and disordered media is presently one of the fastest developing subjects in the theory of random walks and classical Brownian motion (Haus and Kehr 1987; Havlin and Daniel 1987; Bouchaud and Georges Nov. 1990; Klafter and Sokolov 2011), and finds important applications in various areas of science. There is a considerable interest in the studies of various forms of anomalous diffusion and non-ergodicity (Metzler and Klafter 2004; Klafter and Sokolov 2011; Höfling and Franosch 2013; Metzler 2014), based either on the theory of heavy-tailed continuous-time random walk (Montroll and George 1965; Scher and ElliottW 1975), or on models characterized by a diffusivity (i.e. a diffusion coefficient) that is inhomogeneous in time (Saxton 1993) or space (Hottovy et al. 2012a; Cherstvy and Metzler 2013). Particularly impressive is the recent progress in single particle imaging, for instance in biophotonics (cf. Tolić-Nørrelykke et al. 2004; Golding and Edward 2006; Jeon et al. 2011; Weigel et al. 2011; Kusumi et al. 2012; Bakker et al. 2012; Cisse et al. 2013 and references therein), where the single particle trajectories of a receptor on a cell membrane can be traced. It is presently intensively investigated how random walk and classical Brownian motion models with inhomogeneous diffusion may be employed in the description of such phenomena (Massignan 2014; Manzo et al. 2014; Gil et al. 2017).

The examples mentioned above are strictly classical, but the recent unprecedented progress in control, detection and manipulation of ultracold atoms and ions (Lewenstein et al. 2012) are giving us the possibility to perform similar kind of experiments (e.g., single particle tracking to monitor the real time dynamics of given atoms) in the quantum regime (Krinner et al. 2013). Note that such experiments were unthinkable 20 years ago (see the corresponding paragraphs about difficulties to observe QBM in Gardiner and Zoller 2004). Note also that ultracold set-ups will naturally involve spatial inhomogeneities, due to the necessary presence of trapping potentials and eventual stray fields. This is in fact one of the motivation of this work: to formulate and study theory of the QBM in the presence of spatially inhomogeneous damping and diffusion.

An immediate application of our theory concerns ultracold quantum gases. Quantum gases have sparked off intense scientific interest in recent years, both from the theoretical and experimental point of view. They are an excellent testbed for manybody theory, and are particularly useful to investigate strongly coupled and correlated regimes, which remain hard to reach in the solid state field (Bloch et al. 2008; Lewenstein et al. 2012).

In particular QBM may be useful to approach the polaron problem. The concept of *polaron* was introduced in the first part of the last century by Landau and Pekar to describe the behavior of an electron in a dielectric crystal (Landau and Pekar 1948). The motion of the electron distorts the spatial configuration of the surrounding ions, which let their equilibrium positions to screen its charge. The movement of the ions is associated to phonon excitations that dress the electron. The resulting system, which consists of the electron and its surrounding phonon cloud, is called polaron. The concept of polaron has been extended to describe a generic particle, the impurity, in a generic material, e.g. a conductor, a semiconductor or a gas (Fröhlich 1954; Alexandrov and Devreese 2009). One important example is that of an impurity embedded in an ultracold gas. This system has been widely studied both theoretically and experimentally, in the case of a ultracold Fermi (Schirotzek et al. 2009; Kohstall et al. 2012; Koschorreck et al. 2012; Massignan et al. 2014; Lan and Lobo 2014; Levinsen et al. 2014; Schmidt et al. 2012) or Bose gas (Côté et al. 2002; Massignan and Smith 2005; Cucchietti and Timmermans 2006; Palzer et al. 2009; Catani et al. 2012; Spethmann et al. 2012; Richard and Schmidt 2013; Fukuhara et al. 2013; Shashi et al. 2014; Benjamin et al. 2014; Grusdt et al. 2014a, b; Christensen et al. 2015; Levinsen et al. 2015; Ardila et al. 2015; Volosniev et al. 2015; Grusdt and Demler 2016; Grusdt and Fleischhauer 2016; Shchadilova et al. 2016b, a; Castelnovo et al. 2016; Ardila ct al. 2016; Robinson et al. 2016; Jørgensen et al. 2016; Hu et al. 2016; Rentrop et al. 2016).

In the QBM framework, the impurity plays the role of the Brownian particle, while the bath consists of the degrees of freedom related to the gas. The main reason to study this system from the open quantum systems point of view lies in the possibility to better describe the motion of the impurity, rather than its spectral quantities, such as ground state, energy levels and so on, like in the majority of the literature nowadays. The interest in the motion of the impurity is motivated by a recent class of experiments aimed to measure observable related to the impurity dynamics, for instance that of Catani et al. (2012). Here, the physics of an impurity in a gas in one dimension is considered, and its position variance is measured, evaluating in a quantitative manner important features of the motion, such as oscillations, damping and slope. To evaluate this kind of behavior a continuous-variable model such as QBM is appropriate. We shall show that the application of such a model to the *Bose polaron* system (an impurity in a Bose gas) requires the non-linear coupling extension mentioned above.

The book is organized as follows. In Chap. 2 we resume the essentials of classical Brownian motion. This part of the thesis does not contain any original result and basically relies on the material presented in the work of Mazo (2002), but it is important to present the main results of the classical Brownian motion in order to make the manuscript self-consistent. We describe the experimental observation of Robert Brown and then we proceed by going through the theoretical study of Einstein, who wrote an equation for density probability of the pollen grains, termed Fokker-Planck equation. In this way he computed the mean square displacement of the pollen grains, predicting a linear dependence on time (diffusion effect). Actually Einstein was not the only one who tried to propose a theoretical explanation of Brownian motion. Other scientists, such as Marian Von Smoluchowski and then Paul Langevin, dealt with the same problem, although with different techniques. In particular, Langevin treated Brownian motion by means of a stochastic differential equation ruling the temporal evolution of the grains position. He also found diffusion effect for the mean square displacement. This was detected in experiments in 1909 by Perrin, confirming theories of Brownian motion and providing a convincing evidence of the corpuscular essence of the matter.

In Chap. 3 we start to study QBM. The physics of QBM may be explored by means of different formal tools. Among these, the most common is the master equation, i.e. an equation ruling the temporal evolution of the reduced density matrix of the central system, here represented by the quantum Brownian particle. The master equation is a fundamental object in the field of open quantum systems and permits to evaluate in a quantitative manner both decoherence and dissipation, as well as the average values of the observables. However, in many cases the structure of a master equation may result complicated and the procedure to solve it is often not so easy. Therefore, one usually looks into a particular class of approximated master equations, allowing to deal with a certain problem in a mathematical simple manner. An important example is provided by the Born-Markov master equation, based on the absence of self-correlations within the environment (Markov approximation) and the assumption that the global state of the system plus the bath remains separable at all times (Born approximation). In the majority of the situations this kind of equations can be solved analytically providing a description of the behavior of the central system. Comparisons with experiments suggested that the predictions of this model are reasonable in many cases (Moy et al. 1999; Kirton et al. 2012). We derive the Born-Markov master equations for QBM, constituting the quantum analogue of the equation derived by Einstein. We solve this master equation and we focus on its stationary solution, which can be represented in the phase-space by means of a Gaussian Wigner function. We study how its geometrical configuration in the phase-space varies as the system parameters of the system, such as temperature and interaction strength, change: In particular, as the temperature decreases and the interaction strength grows, the quantum Brownian particle experiences cooling and genuine position squeezing. The latter is particularly important: it occurs when the position variance of the particle takes a values smaller than that associated by the Heisenberg principle, although this is fulfilled. Thus it corresponds to high spatial localization of the Brownian particle, namely to a good knowledge of the particle position, compared with the characteristic length scales of the systems.

In Chap. 4 we translate the same analysis to the QBM with a non-linear coupling. We exhibits the Born-Markov master equation for the most general non-linear coupling, paying special attention to the situation where the dependence on the Brownian particle position is quadratic. Here, the Gaussian Wigner function just provides an approximation for the stationary state. Also in this case, it is possible to detect squeezing and cooling as the temperature approaches very low values and the bath-system coupling gets more and more strong. In this regime, anyway, both Born and Markov approximations are not fulfilled and the resulting master equation is not appropriate. In particular it yields to violations of the positivity of the density operator associated to the state of the central system, related to violations of the Heisenberg principle. There exist several methods to overcame this problem. For instance, one could recall a master equation in a Lindblad form. This class of equations was proposed in 1976 in parallel and independently by both Lindblad (1976a) and Gorini et al. (1976).[2] Lindblad master equations arise from the requirement that the positivity of the reduced density matrix is ensured at all times. This type of equations are currently used, for instance to approach the dynamics of the spin-boson model (Leggett et al. 1987). In Chap. 5 we aim to treat the QBM model by means of a Lindblad master equation, in both the case of a linear and non-linear coupling. We find that the Lindblad character of this equation induces a rotation in the phase space of its stationary solution, depending, of course, on the system parameters. Also in this case the stationary state exhibits genuine position squeezing and cooling as the interaction strength increases. When the coupling is non linear, anyway, we find that up a certain value of the system-bath coupling the quantum Brownian particle does not approach a Gaussian stationary state.

We conclude our brief dissertation about QBM by looking into Heisenberg equations. These reduce to an equation for the particle position manifesting the same form of stochastic equation derived by Langevin in a classical context. This quantum Langevin equation carries in general an amount of memory effects according to the properties of the environment. We discuss such a point in detail showing how the presence of memory effects affects the solution of such en equation. In general we will focus on the position and momentum variances, paying particular attention to the low-temperature behavior.

Acknowledgements We acknowledge the Spanish Ministry MINECO (National Plan 15 Grant: FISICATEAMO No. FIS2016-79508-P, SEVERO OCHOA No. SEV-2015-0522, FPI), European Social Fund, Fundació Cellex, Generalitat de Catalunya (AGAUR Grant No. 2017 SGR 1341 and CERCA/Program), ERC AdG OSYRIS and NOQIA, EU FETPRO QUIC, and the National Science Centre, Poland-Symfonia Grant No. 2016/20/W/ST4/00314.

---

[2]Gorini, Kossakowski and Sudarshan submitted their paper on March 19th 1975, and Lindblad one on 7th April 1975, about three weeks later. The former was published in May 1976, while the latter in June 1976.

Chapter 2
Classical Brownian Motion

Brownian motion is the random movement of a particle suspended in a fluid. This phenomenon played a very important role in the history of science because it leaded to the idea that matter is made up by atoms. In this chapter we briefly present the fundamental results concerning classical Brownian motion, focusing on a description of the original observations and of the main theoretical attempts to study it. This part does not contain any original result and it is basically based on the material published in the book of Mazo (2002). However, it is important to address this topic in detail because it yields a physical insight concerning the main features of the phenomenon, such as diffusion, arising also in its quantum counterpart. Moreover, a dissertation on the classical Brownian motion is fundamental in order to make self-consistent the discussion of its quantum extension, on which the thesis relies.

We start by going through the experiment of Robert Brown (Sect. 2.2), who first detected the motion of the pollen grain in a fluid. In addition to observe the movement, Brown recognized that it was not produced by the living origin particle, but it was a matter of dynamics. On this trail, a long time later, the first theoretical attempt to analyze in a quantitative manner Brown's experiment arrived. The author of this study was Albert Einstein, who in 1905 constructed a statistical theory showing that the random movement of pollen grains was due to fluctuations of particles constituting the fluid (Sect. 2.3). This was also argued by the polish Marian von Smoluchowski, who developed a kinetic model to explain Brownian motion in terms of the collisions of the constituents of the fluid embedding the pollen grains. Precisely, both Einstein and Smoluchovski computed the mean square displacement, predicting that it linearly depends on time. Such a behavior is termed diffusion and it has been derived also in another theory created a few years later by Langevin (1908), who proposed to describe Brownian motion by means of a stochastic differential equation (Sect. 2.4). The diffusion behavior has been detected experimentally by Perrin in 1908 and represents a strong confirmation of the theories of Brownian motion (Sect. 2.5). Thanks to this result Perrin won the Nobe prize in 1909 providing a strong evidence for the atomist hypothesis of the matter.

© The Author(s), under exclusive license to Springer Nature Switzerland AG 2019
A. Lampo et al., *Quantum Brownian Motion Revisited*, SpringerBriefs
in Physics, https://doi.org/10.1007/978-3-030-16804-9_2

2.1 Historical Background

In the year 1803, Napoleon sold France's North American colonies to the newborn
United States, for the small sum of sixty million francs. In the American history
books, such a deal is known as "the Luisiana Purchease". President Thomas Jefferson,
wishing to find out exactly what he had bought, sent out an expedition of exploration
under the leadership of Meriweather Lewis and William Clark, which left in 1804,
reached the Pacific Ocean in November 1805, and returned in 1806. Among the
contributions of this mission, one of the least significant would have the greatest
impact, albeit indirectly, on science.

The story of the expedition, based on Lewis' and Clark's journal, constitutes a
fascinating adventure story. In addition to geographical and ethnographical informa-
tions, the expedition also brought back botanical specimens. A genus of plants from
among these specimens, a wild flower found in the Pacific Northwest of the United
States, was named Clarkia Pulchella in honor of Captain William Clark (see Fig.
2.1). In 1826, specimens of Clarkia Pulchella where brought to England by the Scots
botanists David Douglas.

By the year 1827, Robert Brown (1773–1858) was a renewed botanist. As a young
man, Brown studied medicine at Edinburgh, but never finished his studies nor took
a degree. He enlisted in a newly raised Scottish regiment and was posted to Ireland,
where he was appointed Surgeon's Mate, although seems to have spent more time
collecting botanical specimens than attending to patients.

Brown acquired some reputation as a botanist, and he had come to the attention
of Sir Joseph Banks who was organizing an expedition to Australia, or, as it was then
called, New Holland. Banks had need of a botanist for the expedition, and offered
the position to Brown; Brown's medical experience no doubt weighed in his favor.
Robert Brown accepted as soon as he could in spite of his connection with the army,
and his formal career as a botanist begun.

Besides collecting and classifying, Brown made several important discoveries in
botany. Perhaps, the one most celebrated by biologists is the achievement concerning

Fig. 2.1 Clarkia Pulchella's
picture from Wikipedia

the eukaryotic character of the plant cell, namely that they have nucleus. Among the physicians and mathematicians, anyway, he is known primarily for the eponymous motion associated with his name.

2.2 The Brown Experiment

In 1827, Brown was investigating the way in which pollen acted during impregnation. He wanted to employ non-spherical grains, in order to be able to observe their orientation. The first plant he studied under the microscope was just Clarkia Pulchella, whose pollen contains granules varying from about five to six microns in linear dimension. It is these granules, not the whole pollen grains, upon which Brown made his observations. Precisely, he detected the motion of the particles immersed in water, and after frequently repeated observations he concluded that such a motion arose neither from currents in the fluid, nor from its gradual evaporation, but belonged to the particle itself.

This inherent incessant motion of a small particles suspended in a fluid (see Fig. 2.2) is nowadays called Brownian motion in honor of Robert Brown. Although similar observations had been made earlier by other scientists, Brown was the first to treat the phenomenon in a quantitative manner, showing that the motion was not due to the living origin of the particles: it is not a biological phenomenon, but a physical one. For instance, Brown had strongly illuminated the specimens under his microscope and hence had heated them. This caused evaporation of the ambient fluid, and Brown asked whether this evaporation might be causing the motion he observed. To answer this experimentally, he made a mixture of water containing particles with an immiscible oil and shook the mixture; small drops of water were formed in the oil, some containing only a single particle. These were stable and did not evaporate for some time. He realized that, in all the drops formed, the motion of the particles

Fig. 2.2 Simulations of Brownian motion

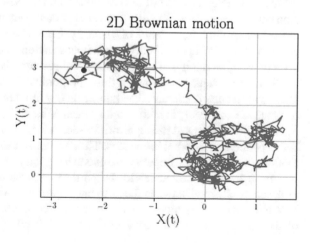

takes place with undiminished activity, although the principal causes assigned for the motion, namely, evaporation and their mutual attraction and repulsion are either materially reduced, or absolutely nil.

Brown had his results printed in a pamphlet, entitled "A brief account of micro-scopical observations made in the months of June, July, and August, 1827 on the particles contained in the pollen of plants; and on the general existence of active molecules in organic and inorganic bodies". This work was originally intended for private circulation but was reprinted in the archival literature shortly after its appear-ance (Brown 1828). Brown used the word "molecule" in the title in a sense different from its current one. It referred to earlier teaching of the Comte de Buffon who introduced this word for the ultimate constituents of the bodies of living beings. This had nothing to do with the later development of Dalton's atomic theory in which the word molecule took on its modern meaning. Brown published a second paper on the motion in the 1829, where he reported the experiments on the oil-water emul-sion mentioned above, and discussed previous observations which could have been interpreted as prior to his (Brown 1829).

After Brown's experiment, thousand of similar systems, constituted by particles suspended in several kinds of environments, have been examined and the motion still manifested the typical character detected by Brown. An important feature of the Brownian motion is: The rapidity of the motions are greater, the smaller the size of the suspended particles (note that the velocity of the Brownian particle does not represent a measurable quantity, this point will be clarified in the next sections). Another important property of the motion is its stability in time. The motion persists as long as the particle remains suspended in the fluid. This has been observed in preparations allowed to stand for over a year. Finally, a very characteristic property is the independence of most external influences. Electric fields, light (as long as it is not absorbed and does not heat the system), gravity (as long as the particles do not settle out) and similar disturbances from outside seem to have no effect. Moreover, the motion exhibits a dependence on the nature of the fluid medium, and especially on its viscosity. Temperature also has a marked effect, however. This could be expected because of the dependence on the viscosity, which is appreciably temperature dependent. Whether there is a residual temperature effect, above that due to the temperature dependence of viscosity, cannot be ascertained on experimental grounds alone. Without a theory to tell how to determine the effect due to viscosity it is not possible see if there is additional temperature dependence.

What is a mechanism showing these properties? The first answer that comes to mind is molecular collisions. The kinetic theory of matter asserts that the molecules of a fluid are constantly in motion with a mean kinetic energy proportional to the tem-perature. For systems at the equilibrium such a kinetic motion is stable in time, and is independent of external influences. Thus we conjecture that the observed irregular motions of a suspended impurity are due to irregular transfers of energy and momen-tum from the fluid molecules to the particle because of the irregularly occurring molecular collisions between the suspended particle and the medium constituents.

There is a nuance which deserves to be clarified. Collisions increasing the velocity of the heavy particle will, on average, be balanced by collisions decreasing that velocity. Thus, the net mean change in velocity due to many collisions will be much smaller than that due to an individual collision. How then can the observed motion be due to collisions? Although the average effect of the collisions will indeed be small, there will be fluctuations about that average. Fluctuations large enough to lead to observable effects, while relatively rare, are still common enough to explain the phenomenon. Precisely, as explained in Sec. 4.1 of the book of Mazo (2002), the fluctuations in velocity due to fluctuations in collision numbers can therefore explain the observed Brownian motion qualitatively.

2.3 The Einstein Theory

We now start to discuss the quantitative attempts to treat Brownian motion. Chronologically, the first relevant one was that of Albert Einstein.[1] who published in 1905 a paper (Einstein 1905) where he proposed an explanation of the behavior observed by Robert Brown.[2] The theory developed by Einstein may be defined as a statistical one, namely it does not rely on a microscopic kinetic model, but it refers to generic probability distributions. This is both its strength and weakness. The strength is that it is applicable to a wide range of circumstances and is easily generalizable. The weakness is that it does not carry so much insight on what is happening at the microscopic dynamical time scale.

In addition to its statistical character, we may define Einstein's theory as a mesoscopic one: it refers to timescales long enough to contain many elementary events, yet short enough to be effectively infinitesimal on an observational scale. Precisely, we introduce a characteristic time τ_r, short compared to macroscopically observable times, yet long with respect to the inverse collision frequency, such that the particle's motions in two consecutive time intervals of length τ_r are independent. Equivalently, we state that successive collisions of the Brownian particle with the surrounding degrees of freedom are independent events. Physically, it means that after a collision between the Brownian particle and a constituent of the environment, the latter interacts with a large number of other constituents, in order to make its dynamical state scarcely dependent on its state before the previous collision with the Brownian one. This happens, for instance, when the density of the medium is sufficiently low so that a collision between the Brownian particle and a given constituent of the medium will be followed by a high number of collisions between the former and other particles of the medium, before it scatters again with the original constituent. In conclusion we may say that during τ_r a large number of collisions occur, in order to destroy the suspended Brownian particle dependence on its initial conditions. In other words we

---

[1]Several studies on Brownian motion have been performed already before of Einstein's, but they do not lead to brilliant conclusions. These works are mentioned in Sect. 1.2 of Mazo (2002).

[2]Einstein also published a second paper (Einstein 1906a). Apparently, in the period between the two publications he had apparently been convinced of the relevance of his considerations for the understanding of the phenomenon associated with the name of Brown. Einstein published two additional short papers on Brownian motion (Einstein 1907, 1908).

say that the process described above is Markovian. Of course Einstein never used such a term, because the famous work of Markov (Markov et al. 1907) concerning Markov chains came two years later.

As we stated in the beginning of the section, Einstein's framework is not based on a kinetic model, but employs probability distributions. Let $p(x, t)$ be the probability density that the particle be at position x at time t. There is no external force so the system is homogeneous, i.e. $p(x, t) = p(-x, t)$. Let $\phi(\Delta, \delta_t)$ be the probability of the particle moving a distance Δ in time δt. The hypothesis of a Markov process permits to write a Chapman-Kolmogorov equation for it:

$$p(x, t + \delta_t) = \int p(x - \Delta, t)\phi(\Delta, \delta_t)d\Delta. \tag{2.1}$$

From the Chapman-Kolmogorov equation, we immediately derive the Fokker-Planck one:

$$\frac{\partial p(x, t)}{\partial t} = D\frac{\partial^2 p(x, t)}{\partial x^2}, \tag{2.2}$$

where

$$D = \lim_{t \to 0} \frac{1}{2\delta_t} \int \Delta^2 \phi(\Delta, \delta_t)d\Delta. \tag{2.3}$$

Equation (2.2) is the well known diffusion equation, and D has the physical significance of the self-diffusion coefficient of the Brownian particle. The derivation of such an equation presented here follows Einstein's very close. Of course, he did not use the terms "Chapman-Kolmogorov" and "Fokker-Planck" in his derivation. These equations had not yet been derived by the authors from whom they are named.

Multiplying both sides of Eq. (2.2) by x^2 and integrate over all space, the left-hand side yields by definition

$$\int x^2 \dot{p}(x, t)dx = \frac{\partial\langle x^2 \rangle}{\partial t}, \tag{2.4}$$

while the right-hand side leads to $6D$, because of the Green theorem. In the end one finds

$$\frac{\partial\langle x^2 \rangle}{\partial t} = 6D, \tag{2.5}$$

and consequently

$$\langle x^2 \rangle = 6Dt. \tag{2.6}$$

The constant of integration must vanish because $\langle x^2 \rangle$ has to be equal to zero at $t = 0$. Since the starting time for observation (the time at which we took the particle position to be at the origin) was arbitrary, this result shows that the sample paths of the random motion are differentiable nowhere because $\Delta r \sim t^{1/2}$. This conclusion applies for all times, because the process has independent increments. Of course, the conclusion is absurd when considered as a microscopic description of the path. It is the result of extrapolating a mesoscopic description down to a microscopic level.[3]

We aim now to express the diffusion coefficient D as a function of the system parameters, such as temperature, etc. Consider a suspension of particles in a fluid with a spatially constant external field, for example gravity, imposed upon it. We denote the external force by F, and choose the z axis of the coordinate system in the direction of F. The potential of the external field is $-Fz$. The suspended particle will move in the z direction and attain a terminal velocity, v, given by

$$v = F/\xi, \tag{2.7}$$

in which ξ is the friction constant. The suspension is supposed to be diluted enough so that the individual particles in it do not interact with each other, but only with the constituents of the surrounding medium. The system is bounded in the z direction; the container has a bottom, for instance. Accordingly, the motion will build up a concentration gradient in the z direction, that produces a diffusion current in the opposite direction to the current induced by the external force. Eventually, the concentration gradient will become large enough that two currents will cancel each other, and the system will reach equilibrium. If we denote the local concentration of suspended particles by $n(z)$, then the particle current induced by the external force is nv; that induced by the concentration gradient is $-D\frac{\partial n}{\partial z}$. This quantity is equal, because of the Fick's law, to the flux current. We have:

$$\frac{nF}{\xi} = -D\frac{\partial n}{\partial z}. \tag{2.8}$$

But at equilibrium $n(z)$ is given by the Boltzmann law

$$n(z) \sim \exp[-Fz/k_BT], \tag{2.9}$$

which replaced in Eq. (2.8) leads to

$$D = k_BT/\xi. \tag{2.10}$$

This is the Einstein relation between the friction coefficient and the diffusion one. Einstein actually did not write it down in this form, but immediately assumed that the friction constant was given by Stokes' law:

$$\xi = 6\pi\eta a, \tag{2.11}$$

where η and a are the viscosity and the radius of the suspended particles, respectively. Replacing Eq. (2.11) into Eq. (2.10) one obtains the Stokes-Einstein relationship:

---

[3]This topic was discussed in Einstein (1906b).

$$D = \frac{k_B T}{6\pi\eta a}. \tag{2.12}$$

Accordingly, Einstein's result for the mean square displacement takes the form

$$\langle x^2 \rangle = \frac{k_B T}{\pi\eta a} t. \tag{2.13}$$

One year later, in 1906, Marian von Smoluchowski (1872–1917) proposed another theoretical model (Smoluchowski 1906) to study Brownian motion.[4] His attempt was based on a kinetic microscopic model, providing a good insight of dynamical mechanisms underlying the phenomenon. Just by studying the collisions by means of the conservation of the momentum and energy, and assuming no memory effects, Marian Smoluchowski derived an expression for the mean square displacement. However, his result was larger than the Einstein's by a factor $\sqrt{\frac{32}{27}} \approx 1$. They are remarkably close considering of all the approximations that went into theory. Smoluchowski and many subsequent commentators claim that his result was, in fact, $\sqrt{\frac{64}{32}}$ larger, but this was because of the error of a factor of two in his formula for the friction constant of heavy particle in a dilute gas. His result was actually closer than he thought. It is now universally agreed, and was agreed even by Smoluchowski, that Einstein result is the correct one.

2.4 The Langevin Theory

The Einstein and Smoluchowski theories look very different on the surface. One employs the dynamics of the particle motion, while the other is a purely statistical theory. A link between the two conceptions was provided in 1908 by Paul Langevin (Langevin 1908). A suspended particle in a fluid is acted upon by forces due to the molecules of the solvent. This force may be expressed as a sum of its average value and a fluctuation around such an average value. Langevin's idea was to treat the mean force dynamically and the residual fluctuating part of the force probabilistically.

We assume that the mean force on a particle moving slowly in a viscous medium is given by

$$F_{av} = -\xi v, \tag{2.14}$$

where v is the velocity of the particle relative to the resting fluid. The differences in sign between it and the relation in Eq. (2.7) is due to the fact that the former refers to the velocity induced by an external force, while the latter concerns the force caused by a given velocity. The two are equal in magnitude and opposite in sign.

---

[4]Smoluchowski approached the study of Brownian motion around 1900, but he published his results only in 1906, under the impetus of Einstein's paper.

We shall denote the residual fluctuating force by X. We know little about X in detail, so we shall make only a few statistically hypothesis about its properties. First of all, X is a fluctuation about a mean, so it must, itself, have zero mean

$$\langle X \rangle = 0. \tag{2.15}$$

Second, we assume that X is a stationary process with a very short correlation time:

$$\langle X(t)X(t+s) \rangle = \langle X^2 \rangle \phi(s), \tag{2.16}$$

in which $\phi(s)$ is a function that is very sharply peaked about $s = 0$. The correlation time is also short compared to M/ξ (the only characteristic time of the system) that ϕ may be taken to be a Dirac delta function, i.e. $\phi(s) \sim \delta(s)$. In other word this is the way in which Langevin implemented in his approach Markov approximation. Moreover it is not correlated with the position of the particle at time t, nor with the velocity at any previous time:

$$\langle X(t)x(s) \rangle = 0, \quad \langle X(t)\dot{x}(s) \rangle = 0, \quad t > s. \tag{2.17}$$

Newton's second law of motion for this system reads

$$M\ddot{x} = -\xi\dot{x} + X(t). \tag{2.18}$$

This differential equation cannot be solved in the usual sense because we do not know enough about $X(t)$. Furthermore, in order that its correlation time as short as assumed X must be a fluctuating function. Precisely we require it is a Wigner measure.

We multiply both sides of Eq. (2.18) for $x(t)$ and then we take the mean value of the result. Recalling the average values in Eq. (2.18) we get

$$M \langle x \frac{\partial v}{\partial t} \rangle = -\xi \langle xv \rangle. \tag{2.19}$$

Since $v = \dot{x}$, such an equation may be put in the form

$$\frac{M}{2} \frac{\partial^2 \langle x^2 \rangle}{\partial t^2} + \frac{\xi}{2} \frac{\partial \langle x^2 \rangle}{\partial t} = 2k_B T, \tag{2.20}$$

where we have used the equipartition theorem $\langle v^2 \rangle = 3k_B T/M$. The solution of this equation is very easy to find. Imposing $\langle x^2(0) \rangle = 0$ we have

$$\langle x^2(t) \rangle = \frac{6k_B T}{\xi}t + B\left[\exp(-\xi t/M) - 1\right], \tag{2.21}$$

where B is an integration constant specified by $\langle xv \rangle = 0$ at time zero. Although we do not need to know it, the value of B can be computed to be $6Mk_BT/\xi^2$. Thus, after a time of the order of ξ/M, Einstein's result is valid. For typical situations, this time is indeed quite short, of the order of 10^{-7} s. Consequently, Einstein's result may be considered to be valid for practical purposes for all time.

Why is this result equal to Einstein's for practical purposes only, and not for all time? Einstein worked completely in the configuration space of the Brownian particle; he never introduced the velocity of the particle. Thus he completely neglected the inertia of the particle and the possibility of the persistence of velocity. In other words, the short time, τ, after which the displacements of the particle should be independent, should be longer than M/ξ. Langevin, on the other hand, worked in the particle's phase space and was able to treat the velocity relaxation. Langevin's description is on a finer scale than that of Einstein.

2.5 The Perrin Experiment

The scientific achievements we presented in this chapter had a great impact on the vision of the nature in the beginning of the nineteenth century. How can one explain the incessant movement of the particle detected by Robert Brown, which seems to contradict the second law of thermodynamics? The key of the answer provided by the theoretical studies of Einstein, Smoluchowski and Langevin lies in fluctuations. Then, what is fluctuating? This may be only explained on the basis of particles.

The idea of a corpuscular reality is the most significant contribution of the Brownian motion to the representation of the world in that period. Actually, since the time of John Dalton (1766–1844) the intuition that the matter was made up of elementary particles called atoms, and their unions, now called molecules, took strong hold in the scientific community. Such a hypothesis became particularly popular especially after the First International Chemistry Conference in Karlsruhe in 1860 where Stanislao Cannizzaro showed how the ideas of Amadeo Avogadro could be used to construct a rational table of atomic weights. Even so, there were skeptics. The most prominent of these where Wilhelm Ostwald (1853–1932) and the physicist Ernest Mach (1838–1916), who argued that there was not any experimental proof of the existence of atoms. Exner (1900) and Svedberg (1906) already made quantitative analysis of Brownian motion (Exner 1900; Svedberg 1906), but they did not have Einstein of Smoluchowski results available; the experiments were not suitable for a detailed verification of the theory. This had to wait for the experiments of Jean Perrin (1870–1942), a convinced atomist.

Perrin aimed to study the dependence of the mean square displacement on the radius of the particle. This kind of experiment was not so easy: It was necessary to prepare a monodisperse suspension, not a trivial task. In 1908 Perrin (and his PhD students) prepared a suspension of particles of gamboge of mastic of uniform size and observed the particles with a camera lucida, a device that projects an image on a plane surface suitable for tracing. He made measurements of the displacements

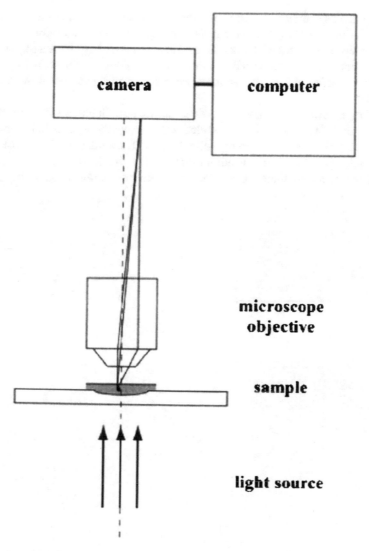

Fig. 2.3 Set-up of the Perrin's experiment revisited by Newburgh et al. (2006)

for as many as 200 distinct granules, confirming the predictions of Einstein and Langevin discussed above (Perrin 1908a, b). Perrin published these results in a long paper (together with his student) in 1909 (Perrin 1908c), and became an energetic proselytizer for the reality of atoms. He received the Nobel Prize in 1926 for his work on the discontinuous structure of matter (Fig. 2.3).

Perrin was apparently also the first to realize that the path of a particle undergoing Brownian motion must have elements in common with a graph of a function which is not differentiable. Such functions had previously been studied by mathematicians and regarded as pathological cases, whose only importance was in illustrating what was really encompassed within the general concept of function. But now, according to Perrin, these so-called pathological functions can be seen to have a physical realization. This idea was taken up by mathematicians and forms the basis for a branch of the theory of stochastic, or random, processes, itself a subfield of the theory of probability.

Perrin's experimental verification of the Einstein-Smoluchowski theory, together with the work of J.J. Thompson on the electron, was rapidly, successful in persuading most of the anti-atomist that atoms really did exist. Ostwald, one of the most prominent skeptics, recanted in a new edition of his influential textbook. Only Mach was not convinced, and continued to consider the existence of atoms as only a hypothesis.

Chapter 3
Quantum Brownian Motion

In the previous chapter we described the main theoretical attempts to treat the grains motion detected by Robert Brown. These theories are purely classical and rely on *phenomenological* equations, i.e. equations that are not derived in a Hamiltonian framework, but are proposed starting from experimental results that one aims to interpret. In this chapter we move our analysis to the quantum domain. The standard procedures of quantization are based on the existence of Hamiltonians (or equivalently Lagrangians) for the system in which one is interested. So, the first step to approach the study of Brownian motion in the quantum regime is to look for a Hamiltonian description of the phenomenon observed by Robert Brown. Precisely, one has to write a Hamiltonian leading to the phenomenological equations, such as those of Einstein [Eq. (2.2)] and Langevin [Eq. (2.18)]. Then, by replacing functional variables with operator ones it is possible to obtain a quantum Hamiltonian for Brownian motion. This is the point of view adopted, for instance, by Caldeira and Leggett (1983a).

In Sect. 3.1 we introduce the Hamiltonian of QBM. It describes a quantum particle, usually trapped in a harmonic potential, coupled to a set of non-interacting harmonic oscillators. The Hamiltonian encodes all the information to study the physics of QBM. There exist several tools to do this. In the current chapter we consider the master equation formalism, i.e. we use an equation for the reduced density operator of the central Brownian particle. Such a topic belongs to standard textbook material, but we present the re-examination published in the paper of Massignan et al. (2015). Here we focused on the evaluation of the stationary state of the system, and the analysis of its geometrical configuration in the phase-space as the system parameters, such as temperature and interaction strength, vary. This kind of analysis leads to the detection of special effects for the impurity, such as squeezing and cooling. We discuss the regime of validity of the method adopted.

© The Author(s), under exclusive license to Springer Nature Switzerland AG 2019   19
A. Lampo et al., *Quantum Brownian Motion Revisited*, SpringerBriefs
in Physics, https://doi.org/10.1007/978-3-030-16804-9_3

3.1 Hamiltonian

The QBM model describes the behavior of a quantum particle interacting with a thermal bath made up by a huge number of harmonics oscillators, satisfying the Bose-Einstein statistics. Despite its simplicity, QBM gained popularity in condensed matter physics due to its very general nature, and its convenience to describe dissipation in a quantum context. The model is defined by the Hamiltonian

$$H = H_S + H_B + H_I, \tag{3.1}$$

where the system, bath and interaction terms are respectively

$$H_S = H_{sys} + V_c(x) = \frac{p^2}{2m} + U(x) + V_c(x), \tag{3.2}$$

$$H_B = \sum_k \left(\frac{P_k^2}{2M_k} + \frac{M_k \omega_k^2 X_k^2}{2} \right) - E_0 = \sum_k \hbar \omega_k a_k^\dagger a_k, \tag{3.3}$$

$$H_I = - \sum_k \kappa_k X_k x. \tag{3.4}$$

In the above expressions p is the particle momentum, m its mass, $U(x)$ the trapping potential depending on its position denoted by x. The expression

$$V_c(x) = \sum_k \frac{\kappa_k^2}{2m_k \omega_k^2} x^2, \tag{3.5}$$

represents the so-called counter-term, needed in the following to remove unphysical divergent renormalizations of the trapping potential arising from the coupling to the bath, as showed in the book of Breuer and Petruccione (2007). The bath bosons have masses M_k and frequencies ω_k, and their momenta and positions are denoted by P_k and X_k, respectively. Alternatively, we describe them with the help of annihilation and creation operators, a_k and $a_k^\dagger$. From the bath Hamiltonian, we have removed the constant zero-point energy E_0.

The parameters in Eq. (3.4) denoted by κ_k characterize the coupling of the bath modes to the system and refer to an interaction term depending linearly on the position of the Brownian particle, as well as on those of the bath constituents. This situation is the conventional one, and corresponds to a quantum system undergoing state-independent damping and diffusion, i.e. damping and diffusion independent on the position (or other observables).

We will restrict our discussion in the following to the one dimensional (1D) case, but generalizations to 2D or 3D are straightforward. Moreover, we shall consider the impurity trapped in a harmonic potential, i.e.

$$U(x) = m\Omega^2 x^2/2, \tag{3.6}$$

where Ω is the frequency of the trap.

3.2 The Born-Markov Master Equation

The Hamiltonian introduced in the previous section embodies all the information one needs to describe the physics of QBM. The high number of constituents of the bath often makes impossible an exact study of the temporal evolution of the whole system. So, one usually proceeds by tracing away the environment's degrees of freedom to focus only on the analysis of the quantum Brownian particle. In this framework, the particle represents a paradigmatic example of open quantum system, namely a system which is not isolated, but continuously affected by the presence of the bath.

There exist several different techniques to handle with open quantum systems and in particular with QBM. In this chapter we shall look into the master equation formalism. A master equation is an equation for the reduced density matrix of the central system and can be considered as the analogue of the Schrödinger equation for open systems. We will study a very special kind of master equation: the Born-Markov master equation, allowing to treat many problems in a mathematically simple form. Comparisons between the predictions of models based on this equation and experiments have shown that the Born and Markov assumptions on which the master equation is based are reasonable in many cases. However we emphasize already at this stage that there are various important physical systems (for example, low-temperature solid-state system) which do not obey to Markovian dynamics and which therefore cannot be appropriately modeled using the Born-Markov master equation. The goal of the present section is the introduction of the general structure of the Born-Markov master equation and the discussion of its regime of validity.

3.2.1 General Structure of a Master Equation

In the ordinary formalism of open quantum systems, the reduced density operator $\rho_S(t)$ is computed via

$$\rho_S(t) = \mathrm{Tr}_E \left[\tilde{U}(t)\rho_{SE}(0)\tilde{U}^\dagger(t) \right], \tag{3.7}$$

where $\tilde{U}(t)$ denotes the time-evolution operator for the whole composite system SE. As is evident from Eq. (3.7), this approach requires that we first determine the state of the total system at a generic instant, before we can arrive at the reduced description through the trace operation. In general this task is not so easy (sometimes impossible) to carry out in practice for the majority of the systems.

In contrast, in the master equation formalism the reduced density matrix $\rho_S(t)$ is calculated directly from an expression of the form

$$\rho_S(t) = \mathcal{L}(t)\rho_S(0), \tag{3.8}$$

where the operator $\mathcal{L}(t)$ is the so-called *dynamical map* ruling the temporal evolution of the central system.[1] Expression in Eq. (3.8) is called *master equation* for $\rho_S(t)$, and it represents the most general structure that such an equation can take.

Obviously, if the master equation is exact, then Eq. (3.7) and (3.8) must be equivalent by definition, i.e. it ensues the identity

$$\mathcal{L}(t)\rho_S(0) = \mathrm{Tr}_E\left[\tilde{U}(t)\rho_{SE}(0)\tilde{U}^\dagger(t)\right], \tag{3.9}$$

and the master equation would amount to nothing else but a trivial rewriting of Eq. (3.7). The master equation approach, thus, is convenient only if one imposes certain assumptions concerning the system-environment states and dynamics, in order to evaluate the approximate time evolution of $\rho_S(t)$, even when it is not possible to compute the exact global dynamics. In fact, here we shall restrict our attention to master equations (valid under particular hypothesis) that may be written as first-order differential equations showing a *local in time* structure, namely which can be cast in the form

$$\dot{\rho}_S(t) = \mathcal{L}[\rho_S(t)] = -\frac{i}{\hbar}[H_S, \rho_S(t)] + \mathcal{D}[\rho_S(t)]. \tag{3.10}$$

This equation is local in time in the sense that the change of the state of the central system at time t depends only on the form of such a state evaluated at t, but not at any other times $s \neq t$. The superoperator $\mathcal{L}$ appearing in Eq. (3.10) acts on $\rho_S(t)$ and typically depends on the initial state of the environment and different terms in the Hamiltonian. To convey the physical intuition behind $\mathcal{L}$, it has been decomposed into two parts:

- A *unitary* part that is provided by the usual von Neumann commutator with the self-Hamiltonian H_S. In general, as we also stated in Sect. 3.1, this term is not identical to the unperturbed free Hamiltonian we indicated by H_{sys}, generating the evolution of the central system in absence of the interaction with the environment. This coupling often perturbs the free Hamiltonian, leading to a renormalization of its spectrum through the introduction of a counter-term, like that in Eq. (3.5). This effect (often termed *Lamb-shift*) has nothing to do with the non-unitary evolution induced by the environment but alters only the unitary part of the reduced dynamics.
- A *non-unitary* part $\mathcal{D}[\rho_S(t)]$ that embodies the action of the environment (decoherence, dissipation and so on). Of course, if such a term is equal to zero, the central system follows an unitary evolution and the resulting master equation differs from

---

[1] Since $\mathcal{L}(t)$ constitutes an operator that in turns acts on another operator, it is commonly named *superoperator*.

the standard Von Neumann one for closed systems only because of the possible presence of a counter-term.

3.2.2 Structure of the Born-Markov Master Equation

The Born-Markov master equation is based on two core approximations that may be stated as below:

- *The Born approximation*, meaning that the coupling between the system and the environment is sufficiently weak and the latter is reasonably large such that changes of its state are negligible and that related to the composite system remains separable at all times, i.e.

$$\rho_S(t) = \rho_S(t) \otimes \rho_E, \tag{3.11}$$

  with ρ_E approximately constant at all times.
- *The Markov approximation*, corresponding to a situation where *memory effects* of the environment are negligible, in the sense that any self-correlation within the environment induced by the interaction with the central system decay rapidly compared to the characteristic timescale over which the state varies noticeably.

Assume now these assumptions hold. Suppose further that the system-bath interaction is described by a Hamiltonian term of the form

$$H_{SE} = \sum_k S_k \otimes E_k, \tag{3.12}$$

where S_k and E_k are self-adjoint operators acting on the Hilbert spaces of the central system and the environment, respectively. Then the evolution of $\rho_S(t)$ is given by *the Born-Markov master equation*

$$\dot{\rho}_S(t) = -\frac{i}{\hbar}[H_S, \rho_S(t)] - \frac{1}{\hbar^2}\sum_k \{[S_k, B_k\rho_S(t)] + [\rho_S(t)C_k, S_k]\}, \tag{3.13}$$

with

$$B_k \equiv \int_0^\infty d\tau \sum_j C_{kj}(\tau) S_j^{(1)}(-\tau), \tag{3.14}$$

$$C_k \equiv \int_0^\infty d\tau \sum_j C_{kj}(-\tau) S_j^{(1)}(-\tau). \tag{3.15}$$

Here $S_j^{(I)}(-\tau)$ denotes the system operator S_j in the interaction picture.[2] The quantity $C_{kj}(\tau)$ is given by

$$C_{kj}(\tau) \equiv \langle E_k^{(I)}(\tau) E_j \rangle_{\rho_E}, \tag{3.16}$$

where the average is taken over the initial environmental state ρ_E (recall that the Born approximation demands that such a state remains approximately constant at all times). The quantity in Eq. (3.16) will be referred to as the *environment self-correlation functions* in the following. The reason for this terminology is easy to understand. The operators E_k can be thought of as observables "measured" on the environment by the interaction with the central system. Self-correlation functions then tell us to what extent the result of such a "measurement" of a particular E_k is correlated with the result of a "measurement" of the same observable carried out a time τ later. Broadly speaking, these functions quantify to what degree the environment retains information over time about its interaction with the system. In fact, the Markov approximation corresponds to the assumption of a rapid decay of these environment self-correlation functions with respect to the timescale set by the evolution of the system. Such a timescale is the relaxation one τ_S, namely that associated to dissipation process, i.e. the transfer of energy from the central system and the environment. Accordingly, we may state that the Markov approximation relies on the following inequality

$$\tau_B \ll \tau_S, \tag{3.17}$$

where the quantity in the left hand-side is the time according which environment self-correlation functions decay. The constraint in Eq. (3.17) is appropriate if the environment is only weakly coupled to the central system, and if the temperature of the bath is sufficiently high.

In conclusion the structure of the Born-Markov master equation of a given system remains fixed by its Hamiltonian and the two approximations discussed above. A clear derivation of Eq. (3.13) goes widely beyond the purpose of the present thesis. However it may be found in Sect. 4.2 of the book of Schlosshauer (2007) and in Sect. 3.3 of the that of Breuer and Petruccione (2007). Moreover, in Sect. 9.1 of their book, Breuer and Petruccione (2007) show that the Born-Markov master equation may be derived even by the Nakajima-Zwanig equation. Precisely, it follows from an expansion in the bath-system coupling constant at the second order. This point of view will be useful to better understand the results we shall present in the chapter.

---

[2]The definition of interaction picture belongs to standard textbook material. A clear explanation may be found in the Appendix of Schlosshauer (2007).

3.3 Born-Markov Equation of Quantum Brownian Motion

In this section we present the Born-Markov master equation of the QBM model, namely we specialize the structure in Eq. (3.13) to the Hamiltonian (3.1). Such a Hamiltonian shows an interaction term in Eq. (3.4) that may be reduced to the decomposition in Eq. (3.12): x plays the role of S_k (with only one index), while $\kappa_k X_k$ is the equivalent of E_k. Thus we can start to evaluate the environment self-correlation function for QBM which in this case remains defined as

$$\mathcal{C}(\tau) = \sum_{ij} \kappa_i \kappa_j \langle X_i^{(\mathrm{I})}(\tau) X_j \rangle_{\rho_{\mathrm{E}}}. \tag{3.18}$$

The terms related to different indexes vanish because of the fact that the environmental oscillators do not interact among them and are therefore completely uncorrelated. Hence, for $i \neq j$,

$$\langle X_i^{(\mathrm{I})}(\tau) X_j \rangle_{\rho_{\mathrm{E}}} = \langle X_i^{(\mathrm{I})}(\tau) \rangle_{\rho_{\mathrm{E}}} \langle X_j \rangle_{\rho_{\mathrm{E}}} = 0, \tag{3.19}$$

since the expectation value of the position coordinate of a harmonic oscillator is equal to zero.

Our task of evaluating $\mathcal{C}(\tau)$ is now reduced to the computation of the averages values referred to equal indexes, i.e.

$$\mathcal{C}(\tau) = \sum_i \kappa_i^2 \langle X_i^{(\mathrm{I})}(\tau) X_i \rangle_{\rho_{\mathrm{E}}}. \tag{3.20}$$

This may be easily accomplished. Let us switch to the representation of the position operators of the bath in terms of the creation and annihilation ones:

$$X_i = \sqrt{\frac{\hbar}{2m_i\omega_i}} \left(a_i + a_i^\dagger \right). \tag{3.21}$$

Then its time evolution in the interaction picture writes as

$$X_i(\tau) = \exp[-\frac{i}{\hbar} H_{\mathrm{E}}\tau] X_i \exp[\frac{i}{\hbar} H_{\mathrm{E}}\tau]$$

$$= \sqrt{\frac{\hbar}{2m_i\omega_i}} \left(a_i e^{-i\omega_i\tau} + a_i^\dagger e^{i\omega_i\tau} \right). \tag{3.22}$$

Accordingly[3]

$$\langle X_i^{(\mathrm{I})}(\tau) X_i \rangle_{\rho_{\mathrm{E}}} = \frac{\hbar}{2m_i\omega_i} \left[\langle a_i a_i^\dagger \rangle_{\rho_{\mathrm{E}}} e^{-i\omega_i\tau} + \langle a_i^\dagger a_i \rangle_{\rho_{\mathrm{E}}} e^{i\omega_i\tau} \right]. \tag{3.23}$$

---

[3]Note that the average values of $a_i a_i$ and its adjoint are zero, as can be easily proved by hand.

But the quantity

$$N_i = \langle a_i^\dagger a_i \rangle_{\rho_{\mathrm{E}}} \tag{3.24}$$

is simply the mean occupation number of the ith oscillator of the environment. By assumption, the environment is in thermal equilibration, which corresponds to assume that

$$N_i \equiv N_i(T) = \frac{1}{\exp[\hbar\omega_i/k_{\mathrm{B}}T] - 1}. \tag{3.25}$$

Using this expression and the standard commutation relation for the creation and annihilation operators it turns

$$
\begin{aligned}
\langle X_i^{(\mathrm{I})}(\tau) X_i \rangle_{\rho_{\mathrm{E}}} &= \frac{\hbar}{2m_i\omega_i} \{ [1 + N_i(T)] e^{-i\omega_i\tau} + N_i(T) e^{i\omega_i\tau} \} \\
&= \frac{\hbar}{2m_i\omega_i} \{ [1 + 2N_i(T)] \cos(\omega_i\tau) - i \sin(\omega_i\tau) \} \qquad (3.26) \\
&= \frac{\hbar}{2m_i\omega_i} \{ \coth\left(\frac{\hbar\omega_i}{2k_{\mathrm{B}}T} \right) \cos(\omega_i\tau) - i \sin(\omega_i\tau) \},
\end{aligned}
$$

where in the last step we employed the fact that

$$
\begin{aligned}
1 + 2N_i(T) &= 1 + \frac{2}{e^{\hbar\omega_i/k_{\mathrm{B}}T} - 1} \\
&= \frac{e^{\hbar\omega_i/k_{\mathrm{B}}T} + 1}{e^{\hbar\omega_i/k_{\mathrm{B}}T} - 1} = \coth\left(\frac{\hbar\omega_i}{2k_{\mathrm{B}}T} \right). \tag{3.27}
\end{aligned}
$$

hence the environment self-correlation function can now be written as

$$
\begin{aligned}
\mathcal{C}(\tau) &= \sum_i \frac{\hbar\kappa_k^2}{2m_i\omega_i} \left[\coth\left(\frac{\hbar\omega_i}{2k_{\mathrm{B}}T} \right) \cos(\omega_i\tau) - i \sin(\omega_i\tau) \right] \\
&\equiv \nu(\tau) - i\eta(\tau), \tag{3.28}
\end{aligned}
$$

Here, the functions

$$
\begin{aligned}
\nu(\tau) &= \frac{1}{2} \sum_i \kappa_i^2 \langle \{ X_i(\tau), X_i \} \rangle_{\rho_{\mathrm{E}}} \\
&= \sum_i \frac{\hbar\kappa_i^2}{2m_i\omega_i} \coth\left(\frac{\hbar\omega_i}{2k_{\mathrm{B}}T} \right) \cos(\omega_i\tau) \\
&\equiv \hbar \int_0^\infty d\omega\, J(\omega) \coth\left(\frac{\hbar\omega}{2k_{\mathrm{B}}T} \right) \cos(\omega\tau), \tag{3.29}
\end{aligned}
$$

$$\eta(\tau) = \frac{1}{2} \sum_i \kappa_i^2 \langle [X_i(\tau), X_i] \rangle_{\rho_E}$$

$$= \sum_i \frac{\hbar \kappa_i^2}{2m_i \omega_i} \sin(\omega_i \tau)$$

$$\equiv \hbar \int_0^\infty d\omega J(\omega) \sin(\omega \tau), \tag{3.30}$$

are commonly named in the literature as *noise kernel* and *dissipation kernel*, respectively.

The function $J(\omega)$, introduced in Eqs. (3.29) and (3.30), is defined as

$$J(\omega) = \sum_i \frac{k_i^2}{2m_i \omega_i} \delta(\omega - \omega_i), \tag{3.31}$$

and is called *spectral density* of the environment. Spectral densities play an immensely important role in the theoretical and experimental study of open quantum systems. They encapsulate the physical properties of the environment once one traces away its degrees of freedom. In modeling the environment, one often goes to a continuum limit in which the description in terms of individual oscillators with discrete frequencies ω_i and masses m_i is replaced by the density $J(\omega)$ corresponding to a continuous spectrum of environmental frequencies ω.

Having successfully determined the environment self-correlation function $\mathcal{C}(\tau)$, we have completed the main step in the derivation of the desired Born-Markov master equation for QBM. The rest of the derivation is now straightforward. The operators in Eqs. (3.14) and (3.15) are immediately written down as

$$B = \int_0^\infty d\tau \mathcal{C}(\tau) x^{(I)}(-\tau), \tag{3.32}$$

$$C = \int_0^\infty d\tau \mathcal{C}(-\tau) x^{(I)}(-\tau), \tag{3.33}$$

where

$$x^{(I)}(\tau) = \exp[-\frac{i}{\hbar} H_S \tau] x \exp[+\frac{i}{\hbar} H_S \tau]$$

$$= x \cos(\Omega \tau) + \frac{p}{m\Omega} \sin(\Omega \tau), \tag{3.34}$$

is the position operator of the quantum Brownian particle in the interaction picture. Inserting Eqs. (3.32) and (3.33) into the general expression for the Born-Markov equation (3.13) we have

$$\dot{\rho}_S(t) = -\frac{i}{\hbar}[H_S, \rho_S(t)] \tag{3.35}$$

$$-\frac{1}{\hbar^2}\int_0^\infty d\tau\{\mathcal{C}(\tau)\left[x, x^{(I)}(-\tau)\rho_S(t)\right] + \mathcal{C}(-\tau)\left[\rho_S(t)x^{(I)}(-\tau), x\right]\}.$$

Recalling the decomposition $\mathcal{C}(\tau) = \nu(\tau) - i\eta(\tau)$ involving the noise and the dissipation kernels and the expression of the position of the Brownian particle in Eq. (3.34), we obtain, rearranging terms properly,

$$\dot{\rho}(t) = -\frac{i}{\hbar}\left[H_S + C_x x^2, \rho(t)\right] - \frac{iC_p}{\hbar m\Omega}[x, \{p, \rho(t)\}] \tag{3.36}$$

$$-\frac{D_x}{\hbar}[x, [x, \rho(t)]] - \frac{D_p}{\hbar m\Omega}[x, [p, \rho(t)]],$$

with

$$C_x = -\int_0^\infty d\tau\,\eta(\tau)\cos(\Omega\tau), \tag{3.37}$$

$$C_p = \int_0^\infty d\tau\,\eta(\tau)\sin(\Omega\tau), \tag{3.38}$$

$$D_x = \int_0^\infty d\tau\,\nu(\tau)\cos(\Omega\tau), \tag{3.39}$$

$$D_p = -\int_0^\infty d\tau\,\nu(\tau)\sin(\Omega\tau). \tag{3.40}$$

The upper limit of the time integrals above is a consequence of the Markovian approximation underlying the derivation of Eq. (3.36). Beyond such an approximation, the upper limit of the integrals is t, rather than ∞, and the coefficients of the master equation get time-dependent. The master equation evaluated with these pre-Markovian coefficients is usually called *Redfield equation*, and constitutes a middle-ground between the Eq. (3.36) and the exact one, valid for arbitrary system-environment interaction strengths. The analytical structure of this time-dependent coefficients have been studied in detail in the literature, for instance by Hu et al. (1992). Of course, this time-dependent coefficients approach the form of the Markovian ones above at long-time.

   Equation (3.36) is the master equation for QBM. It completely rules the dynamics of the central Brownian particle, as well as the decoherence and dissipation processes it undergoes because of the influence of the environment. Moreover, as we will see in the following, it permits to evaluate the observables of the system. Note that we indicated the reduced density matrix of the central Brownian particle by ρ, rather than ρ_S in order to make the notation lighter.

   In this chapter we focus on the case where the spectral density is Ohmic (i.e. it is linear in ω at low frequencies) and has a Lorentz-Drude (LD) cut-off,

$$J(\omega) = \frac{m\gamma}{\pi}\omega\frac{\Lambda^2}{\omega^2 + \Lambda^2}. \tag{3.41}$$

The specific choice of cut-off function yields minor quantitative changes to the coefficients, but as physically expected, it does not alter their asymptotic behaviour. Exploiting the Matsubara representation

$$\coth\left(\frac{\hbar\omega}{2k_B T}\right) = \frac{2k_B T}{\hbar\omega}\sum_{n=-\infty}^{\infty}\frac{1}{1 + (\nu_n/\omega)^2}, \tag{3.42}$$

with frequencies $\nu_n = 2\pi n k_B T/\hbar$, the noise and dissipation kernels may be evaluated analytically with the help of the Cauchy's residue theorem,[4]

$$\nu(\tau) = \frac{m k_B T\gamma\Lambda^2}{\hbar}\sum_{n=-\infty}^{\infty}\frac{\Lambda e^{-\Lambda|\tau|} - |\nu_n|e^{-|\nu_n\tau|}}{\Lambda^2 - \nu_n^2}, \tag{3.43}$$

$$\eta(\tau) = \frac{m\gamma\Lambda^2}{2}\mathrm{sign}(\tau)e^{-\Lambda|\tau|}. \tag{3.44}$$

The explicit form of the dissipation and noise kernels permits to evaluate the analytical structure of the self-correlation function, and in particular its time dependence. We observe that it involves basically the timescales $1/\Lambda$ and $1/\nu_n = \hbar/2\pi n k_B T$ for $n \neq 0$. The largest correlation time is thus equal to

$$\tau_B = \mathrm{Max}\{1/\Lambda, \hbar/2\pi k_B T\}. \tag{3.45}$$

Accordingly the condition in Eq. (3.17) for the applicability of the Born-Markov approximation becomes

$$\hbar\gamma \ll \mathrm{Min}\{\hbar\Lambda, 2\pi k_B T\}, \tag{3.46}$$

where $1/\gamma$ represents the relaxation timescales.

The coefficients can be evaluated as follows:

$$C_x(\Omega) = -\frac{m\gamma}{2\pi}\int_{-\infty}^{\infty}d\omega\,\mathcal{P}\left(\frac{1}{\omega + \Omega}\frac{\omega\Lambda^2}{\omega^2 + \Lambda^2}\right)$$

$$= -\frac{m\gamma\Lambda^3}{2(\Omega^2 + \Lambda^2)}, \tag{3.47}$$

$$C_p(\Omega) = \frac{m\gamma\Omega\Lambda^2}{2(\Omega^2 + \Lambda^2)}, \tag{3.48}$$

$$D_x(\Omega) = \frac{m\gamma\Omega\Lambda^2}{2(\Omega^2 + \Lambda^2)}\coth\left(\frac{\hbar\Omega}{2k_B T}\right). \tag{3.49}$$

---

[4]In particular, the integrals defining both kernels may be reduced to the Fourier transform of a Lorentzian function.

Fig. 3.1 Plots of the adimensional functions $\mathrm{Di}\Gamma(z)$ (continuous) and $\mathrm{Re}[\mathrm{Di}\Gamma[(iz)]]$ (dashed). At large z, both functions approach $\log(z)$ (dotted)

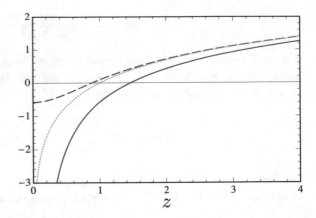

In the first equation above we have used the identity

$$2i \int_0^\infty d\tau \, \sin(\omega\tau) = \int_{-\infty}^\infty d\tau \, \mathrm{sign}(\tau) e^{i\omega\tau} = 2i\mathcal{P}\left(\frac{1}{\omega}\right), \tag{3.50}$$

where $\mathcal{P}$ denotes the principal value of the integral.

The derivation of the anomalous diffusion coefficient D_p is more involved.[5] One has

$$D_p(\Omega) = -\int_{-\infty}^\infty \frac{d\omega}{2\pi} \mathcal{P}\left[\frac{m\gamma\Lambda^2}{\omega+\Omega}\frac{\omega}{\omega^2+\Lambda^2}\coth\left(\frac{\hbar\omega}{2k_BT}\right)\right]. \tag{3.51}$$

To perform the principal part integration with the standard trick

$$\int d\omega \, \mathcal{P}\left[\frac{f(\omega)}{\omega}\right] = \int d\omega \left[\frac{f(\omega)-f(0)}{\omega}\right] \tag{3.52}$$

we need the numerator to be a polynomial in ω. Inserting the Matsubara representation of the hyperbolic cotangent in Eq. (3.42), one finds

$$\frac{\pi(\Omega^2+\Lambda^2)}{m\gamma\Omega\Lambda^2}D_p(\Omega) = -\frac{\pi}{\hbar}\sum_{n=-\infty}^\infty \frac{k_BT}{(\Omega^2+\nu_n^2)}\frac{(\Omega^2-\Lambda|\nu_n|)}{\Lambda+|\nu_n|}$$

$$= \frac{\pi k_BT}{\hbar\Lambda} + \mathrm{Di}\Gamma\left(\frac{\hbar\Lambda}{2\pi k_BT}\right) - \mathrm{Re}\left[\mathrm{Di}\Gamma\left(\frac{i\hbar\Omega}{2\pi k_BT}\right)\right]. \tag{3.53}$$

The function $\mathrm{Di}\Gamma(z) \equiv \Gamma'(z)/\Gamma(z)$ is the logarithmic derivative of the Gamma function, and it is plotted in Fig. 3.1 for both real and imaginary arguments.

---

[5]The name of this coefficient will be explained later.

The C_x term provides a term which strongly renormalizes the harmonic potential frequency. The role of the counterterm V_c introduced in the Hamiltonian is exactly to remove this spurious contribution, and from Eq. (3.36) we see explicitly that a perfect cancellation is obtained by choosing $V_c(x) = -C_x x^2$. Regarding the other coefficients, as we will see in the following, C_p provides momentum damping, D_x yields normal momentum diffusion, and D_p contributes to anomalous diffusion. The D_x term may also be seen as the one responsible for decoherence in the position basis, as widely discussed by Schlosshauer (2007), Zurek (2003), Schlosshauer (2005). There, the density matrix may be represented as $\rho(x_1, x_2, t) = \langle x_1 | \rho(t) | x_2 \rangle$, and one finds $\partial_t \rho(x_1, x_2, t) = -D_x(x_1 - x_2)^2 \rho(x_1, x_2, t)/\hbar + \ldots$, so that the off-diagonal components of ρ decohere at a rate directly proportional to the square of the distance between them, $\gamma_{x_1, x_2}^{(1)} = D_x(x_1 - x_2)^2/\hbar$.

3.3.1 Caldeira-Leggett Limit

We look now into the high-temperature and large cut-off limits

$$k_B T/\hbar \gg \Lambda \gg \Omega. \tag{3.54}$$

Here we may use the series expansions

$$\text{Di}\Gamma(z) = -z^{-1} - \tilde{\gamma} + \pi^2 z/6 + O(z^2) \tag{3.55}$$

$$\text{Re}[\text{Di}\Gamma(iz)] = -\tilde{\gamma} + O(z^2), \tag{3.56}$$

with $\tilde{\gamma}$ the Euler gamma, and real dimensionless argument z, to find

$$\frac{D_p}{\hbar m \Omega} = -\frac{k_B T \gamma}{\hbar^2 \Lambda} + O\left(\frac{\Lambda}{T}\right). \tag{3.57}$$

Replacing it into Eq. (3.36), at high-T one finds

$$\dot{\rho}(t) = -\frac{i}{\hbar}\left[H_{\text{sys}}, \rho(t)\right] - \frac{i\gamma}{2\hbar}[x, \{p, \rho(t)\}] \tag{3.58}$$
$$-\frac{m\gamma k_B T}{\hbar^2}[x, [x, \rho(t)]] + \frac{\gamma k_B T}{\hbar^2 \Lambda}[x, [p, \rho(t)]].$$

Since p is of order $m\Omega x$ in a harmonic potential, the last term may be neglected as it scales as Ω/Λ, and in this way we have the usual *Caldeira-Leggett master equation*:

$$\dot{\rho}(t) = -\frac{i}{\hbar}\left[H_{\text{sys}}, \rho(t)\right] - \frac{i\gamma}{2\hbar}[x, \{p, \rho(t)\}] - \frac{m\gamma k_B T}{\hbar^2}[x, [x, \rho(t)]]. \tag{3.59}$$

Hereafter we will refer to the regime defined by the constraint in Eq. (3.54) as *the Caldeira-Leggett limit*. Note that in the case of a harmonic potential trapping the Brownian particle, or more generally upon neglecting quantum effects for the general non-harmonic potential, the corresponding time dependent equation for the Wigner function in this regime has a particularly simple interpretation (Gardiner and Zoller 2004): it is a Fokker–Plank equation for the probability distribution in the phase space of a classical Brownian particle undergoing damped motion with a damping constant γ under the influence of a Langevin stochastic noise–force $F(t)$. The noise is Gaussian and white, but it fulfills the fluctuation–dissipation relation, i.e. the average of the noise correlation satisfies

$$\langle F(t+\tau)F(t) \rangle = 2\gamma k_B T. \tag{3.60}$$

This relation assures that the stable stationary state of the dynamics is the classical Gibbs-Boltzmann state. In terms of the coefficients entering the master equation the fluctuation–dissipation relation implies that

$$D_x/C_p = 2k_B T/\hbar\Omega. \tag{3.61}$$

3.3.2 Large Cut-Off Limit

We want to look at the limit

$$\Lambda \gg \Omega, k_B T/\hbar, \quad \Omega \sim k_B T/\hbar. \tag{3.62}$$

This is motivated by the fact that, as explained in the context of the applications of QBM to real systems (Lampo et al. 2017, 2018; Charalambous et al. 2019; Mehboudi et al. 2018), the cut-off frequency has a concrete physical meaning and it may be in general much larger than the temperature. In this case we find

$$\dot{\rho}(t) = -\frac{i}{\hbar}\left[H_{\text{sys}}, \rho(t)\right] - \frac{i\gamma}{2\hbar}[x, \{p, \rho(t)\}]$$
$$- \frac{m\gamma\Omega}{2\hbar}\coth\left(\frac{\hbar\Omega}{2k_B T}\right)[x,[x,\rho(t)]] - \frac{D_p}{\hbar m\Omega}[x,[p,\rho(t)]]. \tag{3.63}$$

For large z we have

$$\text{Di}\Gamma(z) \sim \log(z) - 1/(2z) + O(z^{-2}) \tag{3.64}$$

$$\text{Re}[\text{Di}\Gamma(iz)] \sim \log(z) + 1/(12z^2) + O(z^{-3}), \tag{3.65}$$

and the anomalous diffusion coefficient is proportional to

$$D_p \sim \frac{m\gamma\Omega}{\pi} \log\left(\frac{\hbar\Lambda}{2\pi k_B T}\right). \tag{3.66}$$

In this limit, we have moreover

$$D_x/C_p = \coth\left(\hbar\Omega/2k_B T\right). \tag{3.67}$$

Equation (3.63), with the anomalous diffusion coefficient given in Eq. (3.53), constitutes one of the main results of this section.

3.3.3 Ultra-Low Temperature Limit

We finally consider the limit

$$\Lambda \gg \Omega \gg k_B T/\hbar. \tag{3.68}$$

Since both DiΓ functions in Eq. (3.53) diverge logarithmically, the temperature drops completely out of the equations, which reads now

$$\dot{\rho}(t) = -\frac{i}{\hbar}\left[H_{\text{sys}}, \rho(t)\right] - \frac{i\gamma}{2\hbar}[x, \{p, \rho(t)\}] \tag{3.69}$$
$$- \frac{m\gamma\Omega}{2\hbar}[x, [x, \rho(t)]] - \frac{\gamma}{\hbar\pi}\log\left(\frac{\Lambda}{\Omega}\right)[x, [p, \rho(t)]].$$

It is important to remark that the ultra-low temperature limit, as well as that discussed above, have to be considered very carefully. In these regime in fact the Markov approximation could not be properly fulfilled because of the low value of the temperature, as shown in Eq. (3.46).

3.4 Wigner Function Approach

The master equation for the density matrix ρ can be particularly well analyzed in terms of the Wigner function W. The Wigner function is a quasi-probability distribution providing a representation of the density matrix in the phase-space. In order to express Eq. (3.36) in terms of the Wigner function it is useful to introduce the differential operators

$$x_\pm = x \pm \frac{i\hbar}{2}\frac{\partial}{\partial p}, \quad p_\pm = p \pm \frac{i\hbar}{2}\frac{\partial}{\partial x}, \tag{3.70}$$

that satisfy the commutation rules

$$[x_+, x_-] = [p_+, p_-] = 0, \tag{3.71}$$
$$[x_+, p_-] = -[x_-, p_+] = i\hbar.$$

The formal substitutions [see Eqs. (4.5.11) of Gardiner and Zoller 2004] are of great use in the following:

$$x\rho \to x_+ W, \quad p\rho \to p_- W, \tag{3.72}$$
$$\rho x \to x_- W, \quad p \to p_+ W.$$

We note here that, while in the previous Sections x and p stood for the usual non-commuting operators, from now on the same symbols will be used to represent the commuting variables of the Wigner function $W(x, p)$. It turns, for general Ω, Λ and T, the following functional differential equation[6]

$$\dot{W} = \left[m\Omega^2 \partial_p x - \frac{\partial_x p}{m} + \frac{2C_p}{m\Omega} \partial_p p + \hbar D_x \partial_p^2 - \frac{\hbar D_p}{m\Omega} \partial_x \partial_p \right] W. \tag{3.73}$$

This is the equivalent of the Einstein's equation (2.2): the Wigner function plays here the role of the density probability used by Einstein. The fact that the Hamiltonian introduced in the beginning of the chapter permits to recover an equation showing the same form of the classical one derived by Einstein justifies the use of the name "Brownian motion" to indicate the current quantum model. This shall be even clearer in Chap. 6.

The analogy with Einstein's equation allows to better understand the physical meaning of the coefficients (and so their name). For instance it is possible to note that the coefficient D_x multiplies a term yielding a diffusion with respect of the momentum. Similarly, D_p is proportional to mixed diffusion term. For this goal is named *anomalous diffusion term*. The physical meaning of the coefficient will be even clearer when we derive motions equations.

The stationary solution of this equation may be found by inserting a generic Gaussian ansatz

$$W_{st} \propto \exp\left[-\left(\sigma_p \frac{p^2}{2m} + \sigma_x \frac{m\Omega^2 x^2}{2} \right) / (k_B \tilde{T}) \right] \tag{3.74}$$

with real parameters σ_p and σ_x, and equating independently the coefficients of x^2 and p^2 to zero in the resulting equation.

In the oversimplified Caldeira-Leggett limit defined by the constraint in Eq. (3.54), one would set

$$D_x = m\gamma k_B T / \hbar, \quad D_p = 0, \tag{3.75}$$

---

[6]Note that $[p, \rho]\hat{x} = [(p_- - p_+)\rho]\hat{x} = x_-(p_- - p_+)W$.

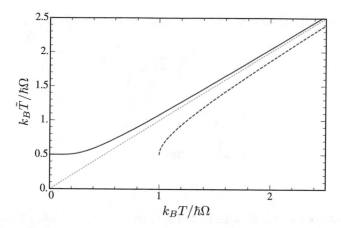

Fig. 3.2 Effective temperatures as obtained through the complete quantum treatment, Eq. (3.78) (blue), and by means of an oversimplified approximation discussed in App. D of Massignan et al. (2015, Eq. (E5)) (red)). The green line is the high-T result, $\tilde{T} = T$

and find in this way

$$\sigma_p = \sigma_x = 1, \quad \tilde{T} = T. \tag{3.76}$$

By retaining instead the complete expression of all terms in the equation (and, in particular, a non-zero D_p), we find that the stationary Wigner function is obtained by choosing $\sigma_p = 1$ and

$$\sigma_x = \frac{1}{1 - 2D_p/(m\Omega^2 \coth[\hbar\Omega/2k_\mathrm{B}T])}, \tag{3.77}$$

yielding an effective temperature

$$\tilde{T} = \frac{\hbar\Omega}{2k_\mathrm{B}} \coth\left(\frac{\hbar\Omega}{2k_\mathrm{B}T}\right). \tag{3.78}$$

This result is shown in Fig. 3.2. A number of interesting conclusions may now be drawn. First of all, a careful treatment of the equation at low-T yields an effective temperature which saturates to the zero-point motion energy. When $\sigma_p = \sigma_x = 1$, the Gaussian stationary solution with an effective temperature $\tilde{T}$ as given by the quantum result in Eq. (3.78) corresponds to the exact quantum thermal Gibbs-Boltzmann density matrix of an harmonic oscillator (the system) at the temperature T. In this case, the contours of the stationary distributions are circles of radius $\sqrt{2k_\mathrm{B}\tilde{T}/\hbar\Omega}$ for arbitrary T (i.e. of radius 1 at $T = 0$).

More generally, in units of the normalized standard deviations

$$\delta_x = 2\sqrt{\frac{m\Omega^2 \langle x^2 \rangle_{\text{st}}}{2\hbar\Omega}} = \sqrt{\frac{2k_B \tilde{T}}{\hbar\Omega\sigma_x}} \tag{3.79}$$

$$\delta_p = 2\sqrt{\frac{\langle p^2 \rangle_{\text{st}}}{2m\hbar\Omega}} = \sqrt{\frac{2k_B \tilde{T}}{\hbar\Omega\sigma_p}}, \tag{3.80}$$

the Heisenberg uncertainty relation requires that

$$\delta_x \delta_p \geq 1, \tag{3.81}$$

i.e. that the contour of the distribution encircles an area not smaller than π. An important effect of D_p is that it allows for a contraction of the distribution in x vs. p. The Heisenberg uncertainty principle then puts an important constraint on our theory, forcing us to exclude the region where the inequality is violated. In Fig. 3.3 we illustrate this region of validity, as obtained by inserting Eq. (3.77) in Eq. (3.81): for any $\Lambda > \Omega$, we find that there exists a critical temperature below which the Heisenberg uncertainty principle is violated. Similar squeezing effects have been discussed in the literature, for instance by Haake and Reibold (1984), in the context of the so called Ullersma model (Ullersma 1966). At $T = 0$, the Heisenberg principle requires $\Lambda < \Omega$.

Interestingly, there are no log-corrections to $\tilde{T}$ coming from the log-divergent term D_p. Such coefficient grows with the cut-off, and at very large values σ_x diverges (i.e. δ_x^2 approaches zero) and becomes negative, yielding a non-normalizable solution. However, this bound always lies beyond the one set by the Heisenberg principle, which requires $\delta_x \delta_p \geq 1$. We may say that the quantum particle immersed in the bath experiences an effective "heating" if the phase-space area encircled by the normalized standard deviations is larger than the one a quantum Gibbs-Boltzmann (GB) distribution would occupy at the same temperature. Since

$$\langle E_k \rangle_{\text{GB}} \langle E_p \rangle_{\text{GB}} = (k_B \tilde{T}/2)^2, \tag{3.82}$$

the system is effectively heated if

$$\delta_x \delta_p > \coth\left(\frac{\hbar\Omega}{2k_B T}\right), \tag{3.83}$$

or equivalently $\sigma_x \sigma_p < 1$. Since $\sigma_p = 1$, this amounts to requiring $D_p < 0$, which remarkably does not depend on γ. Asymptotically, we have

$$k_B T/\hbar > \alpha_{(1)}\Lambda + O(\Omega/T), \tag{3.84}$$

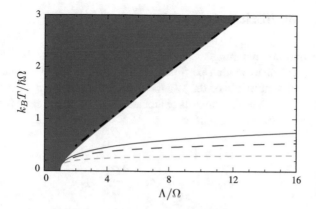

Fig. 3.3 Minimal temperature for the fulfillment of the Heisenberg uncertainty principle for an Ohmic spectral function with Lorentz-Drude cut-off, for $\gamma/\Omega = 0.1$, 0.5, 1 (from bottom to top). In the red region, the gas displays effective "heating" and a quenched aspect ratio in p relative to x (i.e. $\delta_x/\delta_p > 1$). The black, dot-dashed line is the asymptotic approximation to the boundary of unit aspect ratio, $T = \alpha_{(1)}\Lambda$

with $\alpha_{(1)} \approx 0.24$ solution of the implicit equation

$$\pi\alpha_{(1)} + \mathrm{Di}\Gamma(1/2\pi\alpha_{(1)}) + \tilde{\gamma} = 0. \tag{3.85}$$

Finally, we consider the aspect ratio of the phase-space contour described by the standard deviations. Since σ_p always equals unity, it is easy to see that we have a quenched aspect ratio in x, relative to p (i.e. $\delta_x/\delta_p < 1$) in the "cooling" region, and the opposite situation ($\delta_x/\delta_p > 1$) in the "heating" region. In fact the line separating "heating" region from the "cooling" region corresponds to the regime where $D_p = 0$. In this case the Wigner function is exactly given by a Gaussian with effective temperature $\tilde{T}$, and circular shape of the distribution ($\delta_p = \delta_x$); it corresponds precisely to the quantum thermal Gibbs-Boltzmann density matrix.

It should be noted that, when deriving the stationary solutions from a perturbative treatment of the master equation to order $2n$ in the bath-system coupling constant κ_k, one gets a reduced equilibrium state which is exact to order $2n - 2$, and contains some (but not all) terms of the order $2n$ solutions. The overall error is therefore of order $(\kappa_k)^{2n}$ itself, as pointed out by Fleming and Cummings (2011) (for discussion of the nature of exact reduced equilibrium states see also Subaşı et al. 2012). Indeed, the violation of the Heisenberg uncertainty principle we observe within our Born-Markov master equation, which is of second-order in κ_k, is driven by the unphysical logarithmic divergence of D_p, which is itself proportional to γ, i.e. to κ_k^2. Obviously, if the exact master equation is used, then Heisenberg uncertainty violation cannot occur in any parameter regime, ergo this violation is not physical, but is rather a result of applied approximations. On the other hand, it is to be expected that both the degree of cooling and squeezing in the considered quantum stochastic process should be bounded from below, and the Heisenberg uncertainty violation gives a reasonable estimate of this bound.

3.5   Near-Equilibrium Dynamics

We look now into the dynamics of the model in order to get some insight about the motion of the central Brownian particle. For this goal we derive the equations for the first and second moments of the Wigner distribution, that can be easily obtained starting by Eq. (3.73). These moments characterize the Gaussian state fully, and form two closed systems of linear equations:

$$\langle \dot{x} \rangle = \langle p \rangle / m, \tag{3.86}$$

$$\langle \dot{p} \rangle = -m\Omega^2 \langle x \rangle - \frac{2C_p}{m\Omega} \langle p \rangle, \tag{3.87}$$

and

$$\langle \dot{x^2} \rangle = 2\langle xp \rangle / m, \tag{3.88}$$

$$\langle \dot{xp} \rangle = \frac{\langle p^2 \rangle}{m} - m\Omega^2 \langle x^2 \rangle - \frac{2C_p}{m\Omega} \langle xp \rangle - \frac{\hbar D_p}{m\Omega}, \tag{3.89}$$

$$\langle \dot{p^2} \rangle = -2m\Omega^2 \langle xp \rangle - \frac{4C_p}{m\Omega} \langle p^2 \rangle + 2\hbar D_x. \tag{3.90}$$

The solutions of the equations above describe a damped oscillation around their stable stationary values:

$$\langle x \rangle_{st} = \langle p \rangle_{st} = \langle xp \rangle_{st} = 0, \tag{3.91}$$

and

$$\langle p^2 \rangle_{st} = \hbar m\Omega D_x / 2C_p, \tag{3.92}$$

$$(m^2 \Omega^2)\langle x^2 \rangle_{st} = \hbar(m\Omega D_x / 2C_p - D_p / \Omega). \tag{3.93}$$

The only constraint is imposed by the Heisenberg principle

$$\frac{m\Omega^2 \langle x^2 \rangle}{2} \frac{\langle p^2 \rangle}{2m} \geq \left(\frac{\hbar\Omega}{4} \right)^2. \tag{3.94}$$

The equations for $\langle x^2 \rangle_{st}$ and $\langle p^2 \rangle_{st}$ and the resulting Heisenberg bound coincides with the one found for σ_x, σ_p, and $\delta_x \delta_p$ in Sect. 3.4, a fact which should not surprise, as we have seen that a Gaussian Ansatz was providing an exact solution of the problem.

The equations for the first and second moments highlight once more the physical meaning of the coefficients of Eq. (3.36). For instance, Eq. (3.87) shows an exponential decreasing of the average value of the momentum induced by the interaction with environment. Here the coefficient C_p provides an information about the timescale ruling such a process. We see that it depends only on the spectral density, and so its parameters such as cut-off and damping constant, but not on the temperature.

We already discussed in Sect. 3.4 the relation between coefficient D_p and decoherence in the position basis which the particle undergoes. Now, Eq. (3.90) sheds light on its role in the momentum diffusion. Focusing on the time-dependence of $\langle p^2 \rangle$ due only to D_x we obtain

$$\langle p^2 \rangle \propto D_x t, \tag{3.95}$$

manifesting a normal diffusion in the momentum-space. Recall that such normal diffusion is a signature of Brownian motion. In fact, it easy to obtain by Eq. (3.88) that D_x also leads to normal diffusion on the position variance, namely the width of the ensemble in the position space asymptotically grows linearly in time.

Chapter 4
Non-linear Quantum Brownian Motion

The previous chapter was devoted to the discussion of the QBM model. We treated such a model by means of a Born-Markov master equation and we studied its stationary solution. All the theory we developed in Chap. 3 refers to a Hamiltonian model where the interaction depends linearly on the position of the central particle. This is the conventional case, and we will refer to it in the following also as *linear case*. Now we aim to extend our analysis to an interaction term that manifests a non-linear dependence on the variables of the Brownian particle

$$H_{\mathrm{I}} = -\sum_k \kappa_k X_k f(x),$$ (4.1)

The non-linear character of the coupling term above is related to state-dependent damping and diffusion, arising for instance when the quantum Brownian particle is embedded in an inhomogeneous medium, i.e. a medium with a space-dependent density. To keep notation as close as possible to the usual case of linear coupling, we take $f(x)$ to have dimension of length, i.e. we write it as $f(x) = a\tilde{f}(x/a)$, with $\tilde{f}(x)$ being dimensionless, and a denoting a typical length scale on which f varies. In this case the counter-term in Eq. (3.5) writes as

$$V_c(x) = \sum_k \frac{\kappa_k^2}{2m_k\omega_k^2} f(x)^2.$$ (4.2)

In the first part of the chapter we focus on the most simple case provided by an interaction term in the form

$$H_{\mathrm{I}} = -\sum_k \kappa_k X_k \frac{x^2}{a},$$ (4.3)

A. Lampo et al., *Quantum Brownian Motion Revisited*, SpringerBriefs
in Physics, https://doi.org/10.1007/978-3-030-16804-9_4

which we will call *quadratic interaction term*. We derive the Born-Markov master equation which corresponds to the dynamics induced by this Hamiltonian, focusing on the analysis of the stationary solution in the phase-space. Here the Gaussian ansatz just constitutes an approximation. Still, we present a quantitative analysis of the geometrical configuration of the Gaussian Wigner function associated to the stationary state, highlighting the forbidden areas and the range of the values of the system parameters where the quantum Brownian particle experiences squeezing and cooling. In the end, we will characterize the structure of the Born-Markov master equation of the most generic function f in the interaction Hamiltonian in Eq. (4.1). The results we are about to present have been published in the work of Massignan (2015).

4.1 Born-Markov Equation with Quadratic Coupling

In this section we derive the Born-Markov master equation associated to the interaction Hamiltonian term in Eq. (4.3). We note that it differs from the linear case only with respect of the variables related to the central Brownian particle, namely the part associated to the bath operators does not change. Accordingly, the self-correlation function of the environment remains the same, and thus the noise and dissipation kernels, too. Finally, in order to write the Born-Markov master equation we have just to evaluate the time-dependence of the particle position in the interaction picture. It turns:

$$x^2(-\tau) = \left[x \cos(\Omega\tau) - \frac{p}{m\Omega} \sin(\Omega\tau) \right]^2$$
$$= x^2 \cos^2(\Omega\tau) - \frac{\{x, p\}}{m\Omega} \sin(\Omega\tau)\cos(\Omega\tau) + \frac{p^2}{m^2\Omega^2} \sin^2(\Omega\tau), \quad (4.4)$$

so that (using the linearity of commutators and anti-commutators) one finds

$$\dot{\rho}(t) = -\frac{i}{\hbar} [H_S, \rho(t)] - \frac{iC_{xx}}{\hbar a^2} [x^2, \{x^2, \rho(t)\}] - \frac{iC_{xp}}{\hbar a^2} \left[x^2, \left\{ \frac{\{x, p\}}{m\Omega}, \rho(t) \right\} \right]$$
$$- \frac{iC_{pp}}{\hbar a^2} \left[x^2, \left\{ \frac{p^2}{m^2\Omega^2}, \rho(t) \right\} \right] - \frac{D_{xx}}{\hbar a^2} [x^2, [x^2, \rho(t)]]$$
$$- \frac{D_{xp}}{\hbar a^2} \left[x^2, \left[\frac{\{x, p\}}{m\Omega}, \rho(t) \right] \right] - \frac{D_{pp}}{\hbar a^2} \left[x^2, \left[\frac{p^2}{m^2\Omega^2}, \rho(t) \right] \right], \quad (4.5)$$

with the coefficients $C_{...}$ given by

$$C_{xx} = -\int_0^\infty d\tau\, \eta(\tau) \cos^2(\Omega\tau), \tag{4.6}$$

$$C_{xp} = \int_0^\infty d\tau\, \eta(\tau) \sin(\Omega\tau) \cos(\Omega\tau), \tag{4.7}$$

$$C_{pp} = -\int_0^\infty d\tau\, \eta(\tau) \sin^2(\Omega\tau), \tag{4.8}$$

and the $D_{...}$ by

$$D_{xx} = \int_0^\infty d\tau\, \nu(\tau) \cos^2(\Omega\tau), \tag{4.9}$$

$$D_{xp} = -\int_0^\infty d\tau\, \nu(\tau) \sin(\Omega\tau) \cos(\Omega\tau), \tag{4.10}$$

$$D_{pp} = \int_0^\infty d\tau\, \nu(\tau) \sin^2(\Omega\tau). \tag{4.11}$$

Using

$$\sin(x)\cos(x) = \sin(2x)/2, \tag{4.12}$$

and introducing the shorthand notation

$$c(\Lambda) = \Lambda^2/(4\Omega^2 + \Lambda^2), \tag{4.13}$$

for the cut-off function evaluated at frequency 2Ω, we may exploit the results for C_p and D_p of the previous chapter to find

$$C_{xp} = \frac{1}{2}\int_0^\infty d\tau\, \eta(\tau)\sin(2\Omega\tau) = \frac{C_p(2\Omega)}{2} = \frac{m\gamma\Omega}{2}c(\Lambda), \tag{4.14}$$

$$
\begin{aligned}
D_{xp} =& \frac{D_p(2\Omega)}{2} = \frac{m\gamma\Omega}{\pi}c(\Lambda)\{\frac{\pi k_B T}{\hbar\Lambda} + \mathrm{Di}\Gamma\left(\frac{\hbar\Lambda/2}{\pi k_B T}\right)\} \\
& -\frac{m\gamma\Omega}{\pi}c(\Lambda)\mathrm{Re}\left[\mathrm{Di}\Gamma\left(\frac{i\hbar\Omega}{\pi k_B T}\right)\right].
\end{aligned}
\tag{4.15}
$$

Similarly, noting that

$$\cos^2(x) = [1 + \cos(2x)]/2, \tag{4.16}$$

$$I_\nu \equiv \int_0^\infty d\tau\, \nu(\tau) = m k_B T \gamma/\hbar, \tag{4.17}$$

and D_x for the linear case, it turns

$$D_{xx} = \frac{I_v + D_x(2\Omega)}{2} = \frac{m\gamma\Omega}{2}\left[\frac{k_B T}{\hbar\Omega} + c(\Lambda)\coth\left(\frac{\hbar\Omega}{k_B T}\right)\right], \tag{4.18}$$

$$D_{pp} = I_v - D_{xx} = \frac{m\gamma\Omega}{2}\left[\frac{k_B T}{\hbar\Omega} - c(\Lambda)\coth\left(\frac{\hbar\Omega}{k_B T}\right)\right]. \tag{4.19}$$

Finally, recalling

$$I_\eta \equiv \int_0^\infty d\tau\, \eta(\tau) = m\gamma\Lambda/2, \tag{4.20}$$

and the derivation for C_x in Chap. 3, one also obtains

$$C_{xx} = -\frac{I_\eta}{2} + \frac{C_x(2\Omega)}{2} = -\frac{m\gamma\Lambda(2\Omega^2 + \Lambda^2)}{2(4\Omega^2 + \Lambda^2)}, \tag{4.21}$$

$$C_{pp} = -I_\eta - C_{xx} = -\frac{m\gamma\Omega^2}{\Lambda}c(\Lambda). \tag{4.22}$$

In analogy with the linear case, the coefficient C_{xx} diverges with the cut-off Λ, but this poses no problems as

$$[x^2, \{x^2, \rho\}] = [x^4, \rho], \tag{4.23}$$

so this term may always be canceled exactly by an appropriate counter-term

$$V_c(x) = -C_{xx}x^4/a^2, \tag{4.24}$$

representing this time a Lamb-shift of the coefficient of the quartic term in the confinement. All other coefficients remain bounded in the limit of $\hbar\Lambda/k_B T \to \infty$, exception made for D_{xp} which exhibits a mild logarithmic divergence, in complete analogy with D_p in the linear case. The generalized master equation (4.5), together with the explicit forms of its coefficients, represents a central result of the present chapter: it is the main tool to study the dynamics of the central Brownian particle induced by the interaction Hamiltonian (4.3). Here below, we analyze the behavior of the various coefficients in three different limits.

4.1.1   Caldeira-Leggett Limit

In the Caldeira-Leggett regime defined in Eq. (3.54), we have

$$\begin{aligned}
D_{xx} &\approx m\gamma k_B T/\hbar, \\
D_{xp} &\approx -m\gamma(k_B T/\hbar)(\Omega/\Lambda) \longrightarrow 0, \\
D_{pp} &\approx -m\gamma\hbar\Omega^2/(6k_B T) \longrightarrow 0,
\end{aligned} \tag{4.25}$$

and therefore it results

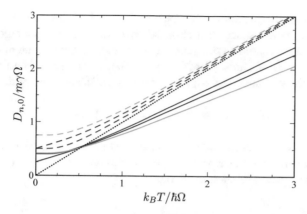

Fig. 4.1 Plot of the coefficients $D_{n,0}$, which control the decoherence rate of the off-diagonal elements of the density matrix $\rho(x_1, x_2)$ in the position basis. The lines represent respectively $D_{1,0} = D_x$ (blue), $D_{2,0} = D_{xx}$ (red), and $D_{3,0}$ (green). Continuous lines are for $\Lambda = 2\Omega$, dashed lines for $\Lambda = 100\Omega$. In the Caldeira-Leggett limit $k_B T/\hbar \gg \Lambda \gg \Omega$, we find $D_{n,0} \to m\gamma k_B T/\hbar$ (dotted line), independent of n

$$\dot{\rho}(t) = -\frac{i}{\hbar}\left[H_{\text{sys}}, \rho(t)\right] - \frac{im\gamma}{2\hbar}\left[\frac{x^2}{a}, \left\{\frac{\{x, p\}}{ma}, \rho(t)\right\}\right]$$
$$- \frac{m\gamma k_B T}{\hbar^2}\left[\frac{x^2}{a}, \left[\frac{x^2}{a}, \rho(t)\right]\right]. \tag{4.26}$$

In this limit, it is easy to identify C_{xp} as being proportional to the momentum damping coefficient, and D_{xx} to the normal momentum diffusion coefficient. In analogy with the traditional QBM model, this latter term may also be seen as the one responsible for decoherence in the position basis. The off-diagonal components of ρ are in this way found to decohere at a rate

$$\gamma_{x_1, x_2}^{(2)} = D_{xx}(x_1^2 - x_2^2)^2/\hbar a^2, \tag{4.27}$$

This is an important result, providing a typical timescale for decoherence of states entangled in position space in presence of a bath coupling of the form $f(x) \propto x^2$. In the end of the chapter we will provide a general formula which yields the position-space decoherence rate $\gamma_{x_1, x_2}^{(n)}$ associated to a coupling with an arbitrary power of the system's coordinate, $f(x) \propto x^n$. Remarkably, and at odds with what found in Hu et al. (1993), we find that superposition states which are symmetric around the origin (e.g., sharply localized around both $+x_0$ and $-x_0$) will be protected by decoherence in presence of couplings containing only even powers of n (Fig. 4.1).

Note also that in this limit we recover again the classical Gibbs-Boltzmann stationary states, and the dynamics satisfies the fluctuation-dissipation relation. Namely, in the case of a harmonic potential, or more generally upon neglecting quantum effects induced by an anharmonic potential, the time dependent equation for the Wigner function has the interpretation of a Fokker–Plank equation for the probability distribution in the phase space of a classical Brownian particle undergoing damped motion with an x–dependent damping $\gamma(x/a)^2$ under the influence of a multiplicative Langevin stochastic noise–force $F(t)(x(t)/a)$. The noise is Gaussian and white, and it fulfills the fluctuation–dissipation relation, i.e. the average of the noise correlation yields

$$\langle F(t+\tau)x(t+\tau)F(t)x(t)\rangle = 2\gamma k_B T \langle x^2 \rangle. \tag{4.28}$$

This relation assures that the stable stationary state of the dynamics is the classical Gibbs-Boltzmann state. In terms of the coefficients entering the master equation the fluctuation–dissipation relation implies that

$$D_{xx}/C_{xp} = 2k_B T/\hbar\Omega. \tag{4.29}$$

4.1.2 Large Cut-Off Limit

Taking the limit

$$\Lambda \gg \Omega, \frac{k_B T}{\hbar}, \tag{4.30}$$

the quantity simply amounts to setting $c(\Lambda) = 1$ in the expression for the various coefficients. Such a regime shows the following features i) the coefficient C_{pp} (a term contributing to a Lamb-shift of the trap frequency Ω) is suppressed as Ω/Λ; ii) the normal momentum diffusion (or position-basis decoherence) coefficient D_{xx}, which is analogous to the D_x of the previous chapter, develops a non-trivial quantum dependence on $\hbar\Omega/k_B T$; iii) the coefficient D_{xp} (which contributes to both the Lamb-shift and the anomalous diffusion) becomes log-divergent in Λ, analogously to D_p for the traditional QBM model; iv) there appears a new coefficient, D_{pp}, which depends on $\hbar\Omega/k_B T$, and vanishes for $k_B T \gg \hbar\Omega$.

We note here that, in this limit, the coefficients of the master equation satisfy the generalized fluctuation-dissipation relations

$$(D_{xx} + D_{pp})/C_{xp} = 2k_B T/\hbar\Omega \tag{4.31}$$

$$(D_{xx} - D_{pp})/C_{xp} = 2\coth(\hbar\Omega/k_B T). \tag{4.32}$$

Finally, we note that the Caldeira-Leggett limit, should be taken with precaution in the case of non-linear coupling. Indeed, as we will see in the following (cf. Fig. 4.2), for strong damping the system in a purely harmonic trap may become dynamically unstable at sufficiently large temperatures.

4.1.3 Ultra-Low Temperature

The master equation for

$$k_B T/\hbar \ll \Omega \ll \Lambda \tag{4.33}$$

reads:

$$
\begin{aligned}
\dot{\rho}(t) = & -\frac{i}{\hbar}\left[H_{\text{sys}}, \rho(t)\right] - \frac{im\gamma}{2\hbar}\left[\frac{x^2}{a}, \left\{\frac{\{x,p\}}{ma}, \rho(t)\right\}\right] \\
& -\frac{m\gamma\Omega}{2\hbar}\left[\frac{x^2}{a}, \left[\frac{x^2}{a} - \frac{p^2}{m^2\Omega^2 a}, \rho(t)\right]\right] \\
& -\frac{m\gamma}{\hbar\pi}\log\left(\frac{\Lambda}{2\Omega}\right)\left[\frac{x^2}{a}, \left[\frac{\{x,p\}}{ma}, \rho(t)\right]\right].
\end{aligned}
\tag{4.34}
$$

As expected the temperature drops out of the equation, and the D_{xp} term is log-divergent in the cut-off Λ. The fact that the self-correlation function writes in the same way for both quadratic and linear QBM implies that the formal condition in Eq. (3.46) associated to the Markov hypothesis shows the same form. According, also in this context the low-temperature limit has to be studied carefully because it could affect the approximations underling the derivation of the Born-Markov treatment, on which discussion relies.

4.2 Stationary Solution

The equation of motion for the Wigner function of a harmonically confined particle reads

$$
\begin{aligned}
\dot{W} = & -\frac{i}{\hbar}\left[\frac{p_-^2 - p_+^2}{2m} + V(x_+) - V(x_-)\right]W \\
& -(x_+^2 - x_-^2)\left[\frac{iC_{xp}\left(\{x_+, p_-\} + \{x_-, p_+\}\right)}{\hbar m\Omega a^2} + \frac{iC_{pp}(p_-^2 + p_+^2)}{\hbar m^2\Omega^2 a^2}\right. \\
& \left. +\frac{D_{xx}(x_+^2 - x_-^2)}{\hbar a^2} + \frac{D_{xp}\left(\{x_+, p_-\} - \{x_-, p_+\}\right)}{\hbar m\Omega a^2} + \frac{D_{pp}\left(p_-^2 - p_+^2\right)}{m^2\Omega^2\hbar a^2}\right]W \tag{4.35} \\
= & \left[-\frac{\partial_x p}{m} + m\Omega^2\partial_p x + \frac{8C_{xp}}{m\Omega a^2}\left(\partial_p p x^2 + \frac{\hbar^2}{4}\partial_p^2(\partial_x x - 1)\right)\right. \\
& +\frac{C_{pp}}{(m\Omega a)^2}\left(4\partial_p x p^2 - \hbar^2\partial_p\partial_x^2 x + 2\hbar^2\partial_p\partial_x\right) \\
& \left. +\frac{4\hbar D_{xx}\partial_p^2 x^2}{a^2} + \frac{4\hbar D_{xp}(\partial_p^2 x p - \partial_p\partial_x x^2 + \partial_p x)}{m\Omega a^2} - \frac{4\hbar D_{pp}(\partial_x x - 1)\partial_p p}{m^2\Omega^2 a^2}\right]W
\end{aligned}
$$

Interestingly, the Gaussian ansatz (3.74) would provide a stationary solution to the equation above if we neglected the terms proportional to C_{pp} and D_{xp}. Remembering that

$$D_{xx} - D_{pp} = 2C_{xp} \coth (\hbar\Omega/k_B T), \tag{4.36}$$

the stationary solution is found when

$$\sigma_p = \sigma_x = 1 \tag{4.37}$$

and

$$k_B \tilde{T} \overset{(C_{pp}=D_{xp}=0)}{=} \frac{\hbar\Omega}{2} \coth \left(\frac{\hbar\Omega}{2k_B T} \right), \tag{4.38}$$

which coincides with the result found in Chap. 3. Unfortunately however D_{xp} is generally not negligible, as for example it diverges logarithmically with the cut-off Λ. In order to incorporate the neglected terms, one may try to generalize the ansatz by including in the exponent terms proportional to higher polynomials in x^2 and p^2 (i.e. terms such as x^4, $x^2 p^2$, or p^4), but no closed solution can be be found in this way, as moments of a given order always couple with higher ones.

The contributions higher than quadratic can, however, be reasonably taken into account by means of the so-called *self-consistent Gaussian (or pairing) approximation* (Gardiner 2009; Risken 2012). The D_{xp} term is proportional to

$$\partial_p^2 xp - \partial_p \partial_x x^2 + \partial_p x \simeq \partial_p^2 \langle xp \rangle_{st} - \partial_p \partial_x \langle x^2 \rangle_{st} + \partial_p x$$

$$= -\partial_p \partial_x \frac{k_B \tilde{T}}{\sigma_x m \Omega^2} + \partial_p x. \tag{4.39}$$

As a general rule, averages of odd functions or partial derivatives vanish when performed with respect to the Gaussian distribution (3.74). Similarly, the C_{pp} term contributes

$$4\partial_p xp^2 - \hbar^2 \partial_p \partial_x^2 x + 2\hbar^2 \partial_p \partial_x \approx \frac{4m k_B \tilde{T}}{\sigma_p} \partial_p x + 2\hbar^2 \partial_p \partial_x, \tag{4.40}$$

as (mixed) derivatives of order higher than two vanish in this approximation. In this way, we get the two equations

$$\delta_p^2 = \frac{\delta_x^2}{\zeta} + \Gamma c_{pp} \left(\frac{\delta_x^2 \delta_p^2}{2} - 1 \right) \tag{4.41}$$

$$\delta_x^2 \delta_p^2 = \frac{\delta_x^2 d_{xx} - \delta_p^2 d_{pp}}{c_{xp}} - 1. \tag{4.42}$$

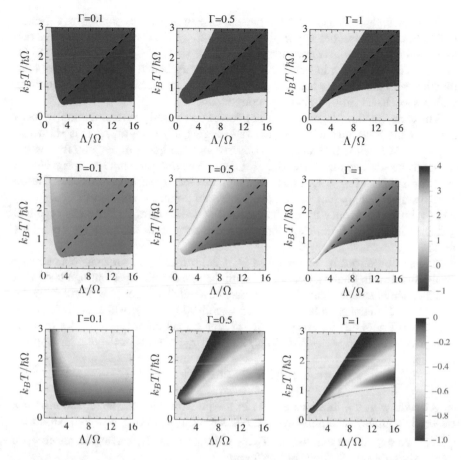

Fig. 4.2 Phase diagram of our equation for a quadratic coupling, under the self-consistent Gaussian approximation. From left to right, plots are for $\Gamma = 0.1$, 0.5, 1. Top **a** the gas experiences an effective "cooling" in the blue regions, and an effective "heating" in the red regions. Center **b** density plot of the logarithm of the aspect ratio $\log(\delta_x^2/\delta_p^2)$. Bottom **c** maximum of the real part of the eigenvalues of the matrix of coefficients of the linear system defined in Eq. (4.54)

To simplify notation, we have introduced the normalized damping $\Gamma \equiv 2\hbar\gamma/(m\Omega^2 a^2)$, the adimensional variables $c_{xp} = 2C_{xp}/(m\gamma\Omega)$ (and similarly for c_{pp}, d_{xp}, ...), and the quantity $\zeta = 1/(1 + 2\Gamma d_{xp})$.

The two coupled Eqs. (4.41) and (4.42) may be combined to obtain a single quadratic equation determining, e.g., δ_x^2, from which we may then extract δ_p^2. The quadratic equation has two solutions, and the correct one may selected by looking at its behaviour in the regime $\Omega \ll k_B T/\hbar \ll \Lambda$. The (-) solution unphysically tends towards zero there. On the other hand, the (+) solution correctly yields $\delta_x^2 \sim 2k_B T/\hbar\Omega$, i.e. an effective temperature $\tilde{T} \sim T$. At odds with the linear case, however, $\tilde{T}$ strongly deviates from T when $T \sim O(\Lambda/\Omega)$.

A detailed phase diagram for the present case of quadratic coupling is presented in Fig. 4.2. The Heisenberg principle requires $\delta_x \delta_p \geq 1$, a condition which gives rise to a minimal acceptable temperature which grows as $T_{min} \sim \log(\Lambda)$ for large Λ/Ω, in close analogy to the linear case. The Heisenberg bound is shown in Fig. 4.2a, together with the region where the gas experiences an effective heating, or cooling, with respect to its Gibbs-Boltzmann counterpart.

The corresponding degree of deformation of the phase-space distribution, as measured by the logarithm of the aspect ratio $\log(\delta_x^2/\delta_p^2) = \log(\sigma_p/\sigma_x)$, is shown in Fig. 4.2b. At small temperatures, we observe the emergence of a region (below the magenta, dot-dashed lines) where $\delta_x^2 < 1$, i.e. of *genuine quantum squeezing*. Notice that, for damping $\Gamma \gtrsim 0.1$, at large temperatures the aspect ratio of the distribution displays a very sharp increase; beyond a certain point, the solution of Eqs. (4.41) and (4.42) yields a value for the fluctuations δ_x^2 which diverges and turns negative, a clearly unphysical feature signaling the breakdown of the Gaussian Ansatz in that region.

It may be noticed by comparing Fig. 4.2a, b that, as in previous chapter, the Gibbs-Boltzmann boundary coincides with the one of unit aspect ratio, a condition which again is independent of Γ. This may be explicitly checked by employing the trial Gibbs-Boltzmann solution $\delta_x^2 = \delta_p^2 = \coth(\hbar\Omega/2k_B T)$, which is an identical solution of Eq. (4.42) for every $\{\Lambda, \Omega, T\}$, and a solution of Eq. (4.41) for every Γ provided that $k_B T/\hbar = \alpha_{(2)}\Lambda + O(\Omega/T)$, with $\alpha_{(2)} \approx 0.189$ satisfying the implicit equation

$$\pi\alpha_{(2)} + 2[\text{Di}\Gamma(1/2\pi\alpha_{(2)}) + \tilde{\gamma}] = 0. \tag{4.43}$$

At odds with the linear case seen in the previous chapter, the equations for a quadratic coupling determine the two ratios $\delta_x^2 \propto \tilde{T}/\sigma_x$ and $\delta_p^2 \propto \tilde{T}/\sigma_p$, but do not provide an explicit expression for $\tilde{T}$, σ_x and σ_p separately, leaving therefore open various possible applications of this theory.

As an example, we may fix $\tilde{T}$ in accordance to the standard formula for the quantum mechanical harmonic oscillator, Eq. (3.78), and then interpret σ_p and σ_x as quantum corrections to the inverse mass $1/m$ and the spring constant $m\Omega^2$. Such "renormalization" should be used if we considered the starting model as a fundamental quantum field theoretic construct.

Alternatively, one may set, say, $\sigma_p = 1$, and consider quantum modification of the effective temperature, and the spring constant. From Eq. (4.41) one finds in this way

$$k_B \tilde{T} = \frac{\hbar\Omega}{2} \frac{\delta_x^2/\zeta - \Gamma c_{pp}}{1 - \Gamma c_{pp}\delta_x^2/2}. \tag{4.44}$$

In the green regions, one of the validity conditions is violated, i.e. either the Heisenberg principle is not satisfied, or one of the eigenvalues of the stability equations becomes positive, or fluctuations δ_x^2 and δ_p^2 are complex numbers. The black dashed lines are the boundaries of unity aspect ratio, where $\delta_x^2 = \delta_p^2$. One needs to examine the nature of Heisenberg uncertainty pathologies in the present quadratic case. Obviously, the exact stationary state should not violate the Heisenberg uncertainty inequality. In the quadratic case, however, the exact solution is not known, and the results of Fleming and Cummings (2011) cannot be applied directly. The pathologies may result from solutions being of mixed order as in Fleming and Cummings (2011), or from the non-Gaussian form of the unknown exact solution. In any case the pathologies signal the invalidity of applied approximations and offer a reasonable bound for the degree of cooling and squeezing in the considered quantum stochastic process.

4.3 Approximated Near-Equilibrium Dynamics

In this case, the Gaussian Ansatz provides only an approximate solution. Again, the first and second moments of the Wigner distribution characterize the Gaussian state fully, but this time they couple to higher moments, so that Wick (Gaussian) de-correlation techniques have to be used. We obtain for the first moments

$$\langle \dot{x} \rangle = \langle p \rangle / m, \tag{4.45}$$

$$\langle \dot{p} \rangle = -m\Omega^2 \langle x \rangle - \frac{8C_{xp}}{m\Omega a^2} \langle x^2 p \rangle - \frac{4C_{pp}}{(m\Omega a)^2} \langle xp^2 \rangle$$
$$- \frac{4\hbar D_{xp}}{m\Omega a^2} \langle x \rangle - \frac{4\hbar D_{pp}}{m^2\Omega^2 a^2} \langle p \rangle.$$

The Wick's theorem allows to replace

$$\langle x^2 p \rangle = \langle x \rangle^2 \langle p \rangle + 2\langle \Delta_x \Delta_p \rangle \langle x \rangle + \langle \Delta_x^2 \rangle \langle p \rangle, \tag{4.46}$$

and similarly for $\langle xp^2 \rangle$, where we represent the Gaussian random variables

$$x = \langle x \rangle + \Delta_x, \quad p = \langle p \rangle + \Delta_p. \tag{4.47}$$

We obtain thus

$$\langle \dot{p} \rangle = -m\Omega^2 \langle x \rangle - \frac{8C_{xp}(\langle x \rangle^2 + \langle \Delta_x^2 \rangle)}{m\Omega a^2} \langle p \rangle$$
$$- \frac{4C_{pp}(\langle p \rangle^2 + \langle \Delta_p^2 \rangle)}{m^2\Omega^2 a^2} \langle x \rangle - \frac{4\hbar D_{xp}}{m\Omega a^2} \langle x \rangle \tag{4.48}$$
$$- \frac{4\hbar D_{pp}}{m^2\Omega^2 a^2} \langle p \rangle - \frac{8\langle \Delta_x \Delta_p \rangle}{m^2\Omega^2 a^2} (C_{pp} \langle p \rangle + 2m\Omega C_{xp} \langle x \rangle).$$

These equations have a stable stationary solution $\langle x \rangle_{st} = \langle p \rangle_{st} = 0$, provided that they describe a damped harmonic oscillator. If such a solution exists, in its vicinity we may identify

$$\langle \Delta_x^2 \rangle_{st} = \langle x^2 \rangle_{st} = \delta_x^2 \hbar/(2m\Omega), \tag{4.49}$$

and

$$\langle \Delta_p^2 \rangle_{st} = \langle p^2 \rangle_{st} = \hbar m\Omega \delta_p^2/2, \tag{4.50}$$

since by hypothesis the first moments are zero, and we may neglect the quadratic terms $\langle x \rangle^2$, $\langle p \rangle^2$ and the crossed fluctuation term $\langle \Delta_x \Delta_p \rangle$, to get the two simultaneous conditions

$$1 + \Gamma d_{xp} + \Gamma c_{pp}\delta_p^2/2 \geq 0, \qquad c_{xp}\delta_x^2 + d_{pp} \geq 0 \tag{4.51}$$

These, in turn, depend self-consistently on the equations for the second moments,

$$\langle \dot{x^2} \rangle = \frac{2}{m}\langle xp \rangle, \tag{4.52}$$

$$\langle \dot{xp} \rangle = \frac{\langle p^2 \rangle}{m} - m\Omega^2 \langle x^2 \rangle - \frac{8}{m\Omega a^2}\left[C_{xp}\langle x^3 p \rangle + \hbar D_{xp}\langle x^2 \rangle \right]$$
$$- \frac{1}{m^2\Omega^2 a^2}\left[C_{pp}\left(4\langle x^2 p^2 \rangle - 2\hbar^2 \right) + 8\hbar D_{pp}\langle xp \rangle \right],$$

$$\langle \dot{p^2} \rangle = -2m\Omega^2 \langle xp \rangle - \frac{4C_{xp}}{m\Omega a^2}\left(4\langle x^2 p^2 \rangle + \hbar^2 \right)$$
$$- \frac{8C_{pp}}{m\Omega a^2}\langle xp^3 \rangle + \frac{8\hbar D_{xx}}{a^2}\langle x^2 \rangle - \frac{8\hbar D_{pp}}{m^2\Omega^2 a^2}\langle p^2 \rangle.$$

From the first equation, we see that if a stable stationary solution exists then $\langle xp \rangle_{st} = 0$. The quartic terms may be decomposed as above, using the Wick's method, and in this way one may compute the stationary solution. A straightforward calculation then shows that in the stationary state described by the momenta $\langle x^2 \rangle_{st}$ and $\langle p^2 \rangle_{st}$ satisfy Eqs. (4.41) and (4.42). To check the stability of the steady-state, we write

$$\langle x^2 \rangle = \langle x^2 \rangle_{st} + \Delta_{x^2}, \quad \langle p^2 \rangle = \langle p^2 \rangle_{st} + \Delta_{p^2}, \quad \langle xp \rangle = \Delta_{xp}, \tag{4.53}$$

and perform linear stability analysis in Δ's,

$$\partial_t(\Delta_{x^2}) = \frac{2}{m}\Delta_{xp} \tag{4.54}$$

$$\partial_t(\Delta_{xp}) = \frac{\Delta_{p^2}}{m} - m\Omega^2 \Delta_{x^2} - \frac{24C_{xp}\langle x^2 \rangle_{st}\Delta_{xp} + 8\hbar D_{xp}\Delta_{x^2}}{m\Omega a^2}$$
$$- \frac{4C_{pp}(\langle x^2 \rangle_{st}\Delta_{p^2} + \langle p^2 \rangle_{st}\Delta_{x^2}) + 8\hbar D_{pp}\Delta_{xp}}{m^2\Omega^2 a^2}$$

$$\partial_t(\Delta_{p^2}) = -2m\Omega^2 \Delta_{xp} - \frac{16C_{xp}}{m\Omega a^2}[\langle p^2 \rangle_{st}\Delta_{x^2} + \langle x^2 \rangle_{st}\Delta_{p^2}]$$
$$- \frac{24C_{pp}}{m^2\Omega^2 a^2}\langle p^2 \rangle_{st}\Delta_{xp} + \frac{8\hbar D_{xx}}{a^2}\Delta_{x^2} - \frac{8\hbar D_{pp}}{m^2\Omega^2 a^2}\Delta_{p^2}.$$

The stability requires that the real parts of all eigenvalues of the matrix governing the above linear evolution have to be negative, i.e. have to describe damping. Numerical analysis of the eigenvalues of this matrix is presented in Fig. 4.2c. The plot indicates that all eigenvalues are negative in most of the region of existence of the physically Gaussian stationary solution, but at the same time the region of validity rapidly shrinks with increasing damping Γ. To summarize, regions colored in green are not accessible by the system because either the normalized standard deviations δ_x^2 and δ_p^2 have an unphysical imaginary part, or they do not satisfy the Heisenberg bound $\delta_x^2 \delta_p^2 \geq 1$, or the equations for the first moments do not describe a damped harmonic oscillator (i.e. inequalities in (4.51) are not satisfied), or at least one of the eigenvalues of the linear stability matrix of the second moments (4.54) becomes positive.

Note that on top of the stability question, Eqs. (4.52) and (4.54) incorporate quantum dynamical effects: they describe dynamics clearly different from their high T classical analogues, due to the quantum form/origin of the diffusion coefficients D_{xx}, D_{xp} and D_{pp}.

Finally, let us comment about the large prohibited region we find in the quadratic case at large T. This region is generally dynamically unstable, and arises because of the diverging fluctuations in x caused by a large Lamb-shift of the effective trap frequency, which turns the attractive harmonic potential into an effectively repulsive one. It is reasonable to expect that this region would become allowed if we added a quartic term to the confinement, on top of the usual quadratic one. Indeed, Hu et al. (1993) considered only this case, for non-linear couplings. However, traps for ultracold atoms are generally to a very high approximation purely quadratic in the region where the atoms are confined, so that the presence of a quartic component may be unjustified in a real experiment.

4.4 General Non-linear Coupling

We consider here the interaction term with a completely general coupling in the position of the particle in Eq. (4.1). If $f \in C^{\infty}(I)$ and thus may be expanded in Taylor series, the master equation can be written in the form:

$$\dot{\rho} = -\frac{i}{\hbar}[H_S, \rho] - \sum_{j,n=0}^{\infty} \sum_{k=0}^{n} \frac{\tilde{f}^{(j)} \tilde{f}^{(n)}}{a^{j+n-2} j! n! (m\Omega)^k} \mathcal{L}_{n,k,j}[x, p, \rho] \qquad (4.55)$$

with

$$\mathcal{L}_{n,k,j}[x, p, \rho] = \left[x^j, \frac{iC_{n,k}}{\hbar} \{\sigma(x^{n-k} p^k), \rho\} + \frac{D_{n,k}}{\hbar} \left[\sigma(x^{n-k} p^k), \rho \right] \right] \qquad (4.56)$$

where $\sigma(x^m p^k)$ is the sum of the $\frac{(m+k)!}{m!k!}$ distinguishable permutations of the $m + k$ operators in the polynomial $x^m p^k$ [e.g., $\sigma(x^2 p) = x^2 p + xpx + px^2$]. We have introduced here

$$C_{n,k}(\Omega) = (-1)^{k+1} \int_0^\infty d\tau \, \eta(\tau) \cos^{n-k}(\xi) \sin^k(\xi) \qquad (4.57)$$

$$D_{n,k}(\Omega) = (-1)^k \int_0^\infty d\tau \, \nu(\tau) \cos^{n-k}(\xi) \sin^k(\xi)$$

where $\xi = \Omega\tau$. These integrals may be calculated by recalling the properties of the Laplace transform. Alternatively, one can employ standard trigonometric identities to straightforwardly reduce every $C_{n,k}$ to a linear combination of C_x and C_p (the ones computed in the linear case), and similarly every $D_{n,k}$ in terms of D_x and D_p. As an example, since

$$\cos^3(\xi) \sin(\xi) = [2\sin(2\xi) + \sin(4\xi)]/8, \qquad (4.58)$$

it is obvious that

$$D_{4,1}(\Omega) = [2D_p(2\Omega) + D_p(4\Omega)]/8, \qquad (4.59)$$

In complete analogy with the linear and quadratic cases, for a power law coupling with $f(x) = a(x/a)^n$ the coefficient $D_{n,0}$ determines the decoherence in the position basis, which for a quantum superposition of two states centered respectively at x and x' happens with a characteristic rate

$$\gamma^{(n)}_{x_1,x_2} = D_{n,0}(x_1^n - x_2^n)^2 / \hbar a^{2n-2}. \qquad (4.60)$$

As a consequence, for an even more general coupling containing various powers of (x/a), the total decay rate in position space reads

$$\gamma_{x_1,x_2} = \sum_{j,n=0}^\infty \frac{\tilde{f}^{(j)} \tilde{f}^{(n)} D_{n,0}(x_1^n - x_2^n)^2}{\hbar j! n! a^{j+n-2}}. \qquad (4.61)$$

In contrast with the work of Hu et al. (1993), we find here that quantum superpositions which are sharply localized at positions symmetric with respect to the origin (e.g., in the vicinity of, say, x_0 and $-x_0$) will be characterized by a vanishing decoherence rate (i.e. a diverging lifetime) in presence of couplings which contain only even powers of n. The decoherence rates in Eq. 4.61 are plotted in Fig. 4.1.

Large cut-off limit
In the limit $\Lambda \gg k_B T/\hbar$, Ω, we find:

- $C_{n,k} \propto \Lambda^{1-k}$, such that at every order n the only divergent term is linear, and it is the one which may be re-absorbed in the Hamiltonian; indeed, $C_{n,0}$ is the coefficient in front of the term $i[x^n, \{x^n, \rho\}] = i[x^{2n}, \rho]$, so that the divergent term is canceled by taking $H_{sys} = H_S - C_{n,0} f(x)^2$. Moreover, for every n we have $C_{n,1} = m\gamma\Omega/2$.

- between the coefficients $D_{n,k}$, only the term with $k = 1$ diverges, logarithmically as $D_{n,1} \sim \frac{m\gamma\Omega}{\pi} \log\left(\frac{\hbar\Lambda}{2\pi k_B T}\right) + \dots$. All terms with $k \neq 1$ are instead finite.

High-temperature limit

In the high-temperature limit $\frac{k_B T}{\hbar} \gg \Lambda \gg \Omega$, the coefficients C are as in the large-cut-off limit, as they do not depend on T. In the set of D coefficients, only $D_{n,0} \sim m\gamma k_B T/\hbar$ remains finite, while all others go to zero. Using the identity $\sigma(x^{n-1}p) = n\{x^{n-1}, p\}/2$, it is easy to show that the master equation (4.55) reduces at high temperatures to

$$\dot{\rho} = -\frac{i}{\hbar}[H_{\text{sys}}, \rho] - \frac{i\gamma m}{2\hbar}[f(x), \{\dot{f}(x), \rho\}] - \frac{m\gamma k_B T}{\hbar^2}[f(x), [f(x), \rho]], \quad (4.62)$$

where

$$\dot{\rho} = -\frac{i}{\hbar}[H_{\text{sys}}, \rho] - \frac{i\gamma m}{2\hbar}[f(x), \{\dot{f}(x), \rho\}] \qquad (4.63)$$
$$- \frac{m\gamma k_B T}{\hbar^2}[f(x), [f(x), \rho]].$$

In this *classical* limit, we see that in presence of a non-linear coupling the coefficients of the QME satisfy a generalized fluctuation-dissipation theorem, since for any n we have $D_{n,0}/C_{n,1} \approx 2k_B T/\hbar\Omega$.

Chapter 5
A Lindblad Model for Quantum Brownian Motion

Chapters 3 and 4 we investigated the QBM model and its extensions by means of Born-Markov master equations. This approach leads to violations of the Heisenberg principle as the temperature decreases and interaction strength grows. Such a pathology avoids the possibility to study the low-temperature regime, as well as that associated to certain values of the system-bath coupling. Overcoming this problem is a fundamental step towards a correct description of the dynamics of the quantum Brownian particle.

There are several possible manners to deal with the violations of the Heisenberg uncertainty principle. First of all, one has to note that, obviously, if the exact master equation is used, violation of Heisenberg principle cannot occur in any parameter regime. Conversely, the master Eq. (3.36) (as well as that in Eq. (4.5)) is the result of a perturbative expansion to the second order in the bath-particle coupling constant (actually, expanding to second order requires weaker assumptions than the Born and Markov ones; the resulting equation may still take into account some non-Markovian effects which vanish in the limit of large times, as shown in the book of Breuer and Petruccione (2007)). In the work of Fleming and Cummings (2011) it has been shown that Heisenberg principle violations in the stationary state disappear if one performs a perturbative expansion beyond the second order in the coupling constant.

In the present chapter we aim to cure the forbidden areas detected in the Born-Markov approach by recalling a Lindblad master equation. Such master equations preserve the positivity of the density operator at all times (Lindblad 1976a; Schlosshauer 2007; Breuer and Petruccione 2007), and this in turn guarantees that the Heisenberg uncertainty principle is always satisfied. A brief, self-contained demonstration of the latter is given in Appendix A. Various ways of addressing this difficulty have been put forward by Lindblad (1976b), Diósi (1993), Isar et al. (1994), Săndulescu and Scutaru (1987), Gao (1997, 1999), Wiseman and Munro (1998), Gao (1998), Ford and O'Connell (1999), Vacchini (2000). We first consider, in Sect. 5.1,

© The Author(s), under exclusive license to Springer Nature Switzerland AG 2019
A. Lampo et al., *Quantum Brownian Motion Revisited*, SpringerBriefs
in Physics, https://doi.org/10.1007/978-3-030-16804-9_5

the master equation (3.36), associated to the Hamiltonian with the linear interaction in Eq. (3.4). We shall refer in the following to this situation as *linear case*. We add a term to the Eq. (3.36), that vanishes in the classical limit, bringing the equation to the Lindblad form and, in particular, ensuring that the Heisenberg principle is always satisfied (Lindblad 1976a). We then rewrite it in the Wigner function representation, deriving the time-dependent equations for the moments of this distribution, showing that they have an exact Gaussian solution, and study in detail its long-time behavior. Up to this point, the results we present belong to well-known papers, such as that of Gao (1997). The original part of the chapter lies in the study of the stationary Gaussian solution in the phase space, and has been published in the paper of Lampo et al. (2016). In particular, we analyze the correlations induced by the environment, which cause a rotation and distortion of the distribution, as well as squeezing effects expressed by the widths and the area of the distribution's effective support.

In the second part of the chapter we move our analysis to the QBM with a quadratic coupling in Eq. (4.3). We again modify the related master equation to obtain a Lindblad one and we study its stationary solutions in the phase space (Wigner) representation. For the quadratic QBM, the exact stationary state is no longer Gaussian, but a Gaussian approximation can be used in certain regimes. However, when the damping is strong, the Gaussian ansatz does not converge for large times, showing that it is not a good approximation to a stationary state.

5.1 Linear Case

A Lindblad master equation has the form

$$\frac{\partial \rho}{\partial t} = -\frac{i}{\hbar}[H_S, \rho] + \sum_{i,j} \kappa_{ij}\left[A_i \rho A_j^\dagger - \frac{1}{2}\{A_i^\dagger A_j, \rho\}\right], \qquad (5.1)$$

where A_i are called Lindblad operators and (κ_{ij}) is a positive-definite matrix. The derivation of Eq. (5.1) goes widely beyond the goal of this thesis, and it may be found anyway in Sect. 3.2.1 of the book of Breuer and Petruccione (2007) or in that of Rivas and Huelga (2012).

Following the approach proposed by Gao (1997) we will replace the Born-Markov master equation (3.36), which cannot be brought to a Lindblad form, by an equation of the form Eq. (5.1) with a single Lindblad operator of the form

$$A_1 = \alpha x + \beta p, \qquad \text{with } \kappa_{11} = 1. \qquad (5.2)$$

Substituting this operator into Eq. (5.1) we obtain

$$\frac{\partial \rho}{\partial t} = -\frac{i}{\hbar} \left[H_S', \rho \right] - i \frac{C_{XP}}{\hbar} \left[x, \{p, \rho\} \right] - \frac{D_{XX}}{2\hbar^2} \left[x, [x, \rho] \right] \tag{5.3}$$

$$- \frac{D_{XP}}{\hbar^2} \left[x, [p, \rho] \right] - \frac{D_{PP}}{2\hbar^2} \left[p, [p, \rho] \right],$$

with

$$H_S' = H_S - \frac{C_{XP}}{2} \{x, p\} \equiv H_S + \Delta H \tag{5.4}$$

and

$$D_{XX} = \hbar^2 |\alpha|^2, \qquad\qquad D_{XP} = \hbar^2 \text{Re} \left({}^* \text{fi} \right), \tag{5.5}$$

$$D_{PP} = \hbar^2 |\beta|^2, \qquad\qquad C_{XP} = \hbar \text{Im} \left({}^* \text{fi} \right).$$

One could obtain the same result employing two Lindblad operators, proportional to x and p respectively. Without loss of generality, we may take α to be a positive real number since multiplying A_1 by a phase factor does not change Eq. (5.1), and we will restrict ourselves to $\text{Im}\beta > 0$, because, as seen from Eq. (5.5), $\alpha \text{Im}(\beta)$ is the damping coefficient C_{XP}, which must be positive.

Equation (5.3) differs from Eq. (3.36) just by two extra terms, involving D_{PP} and ΔH. Equating the coefficients of the remaining terms with those of the analogous terms appearing in Eq. (3.36), one finds

$$D_{XX} = 2\hbar D_x, \qquad\qquad D_{XP} = \frac{\hbar D_p}{m\Omega}, \tag{5.6}$$

$$C_{XP} = \frac{C_p}{m\Omega}, \qquad\qquad D_{PP} = \frac{(\hbar C_{XP})^2 + D_{XP}^2}{D_{XX}}.$$

In the Caldeira-Leggett limit defined in Eq. (3.54), these reduce to

$$C_{XP} \approx \gamma/2, \tag{5.7}$$

$$D_{XX} \approx 2m\gamma k_B T,$$

$$D_{XP} \approx -\gamma \frac{k_B T}{\Lambda},$$

$$D_{PP} \approx \frac{\gamma k_B T}{2m\Lambda^2}.$$

Following Schlosshauer (2007), since the quantities represented by P and $m\Omega X$ have generally the same order of magnitude, one can argue, as in Eq. (5.56) of the book of Schlosshauer (2007), that the terms proportional to D_{XP} and D_{PP} are negligible in the Caldeira-Legget limit, recovering the structure of the usual master equation. We may state so that the Caldeira-Leggett equation (3.59) approximate the Lindblad master equation (5.3). This is in agreement with the fact that at large values of the cut-off and at high-temperature (the Caldeira-Leggett limit) no violations of the Heisenberg principle arise.

The operator ΔH can be absorbed into the unitary part of the dynamics defined by Eq. (5.3), so it can be eliminated by introducing a counter term into the system's Hamiltonian. More generally, we will add to H_S a counter term

$$H_C = (r - 1)\Delta H,\tag{5.8}$$

which depends on a parameter $r \in \mathbb{R}$, leading to the modified Hamiltonian

$$\begin{aligned}H_S' &= H_S - (rC_{XP}/2)\{x, p\}\\ &= \frac{(p - mrC_{XP}x)^2}{2m} + \frac{m(\Omega^2 - r^2C_{XP}^2)x^2}{2}.\end{aligned}\tag{5.9}$$

The effect of r is twofold: it introduces a gauge transformation which shifts the canonical momentum p, and it renormalizes the frequency of the harmonic potential. In the rest of the section we shall study the dynamics defined by Eq. (5.3), first for general values of r and then, for the discussion of the stationary state, focusing on the case $r = 0$. We stress that the introduction of a counter term in the Hamiltonian does not affect the Lindblad character of Eq. (5.3), since it just enters in its unitary part.

5.1.1   Solution of the Lindblad Equation

We are interested in the study of the long-time dynamics of the Brownian particle. In particular, we consider its configuration in the phase space, employing the Wigner function representation. In terms of the Wigner function, Eq. (5.3) becomes $\dot{W} = \mathcal{L}W$, with

$$\begin{aligned}\mathcal{L}W = &-\frac{p}{m}\frac{\partial W}{\partial x} + m\Omega^2 x\frac{\partial W}{\partial p}\\ &+ C_{XP}\left[r\frac{\partial}{\partial x}(xW) + (2 - r)\frac{\partial}{\partial p}(pW)\right]\\ &+ \frac{1}{2}\left[D_{XX}\frac{\partial^2 W}{\partial P^2} + D_{PP}\frac{\partial^2 W}{\partial x^2}\right] - D_{XP}\frac{\partial^2 W}{\partial x\partial p}.\end{aligned}\tag{5.10}$$

Equivalently, one can look at the equations for its moments

$$\frac{\partial \langle x \rangle_t}{\partial t} = \frac{\langle p \rangle_t}{m} - rC_{XP}\langle x \rangle_t \tag{5.11}$$

$$\frac{\partial \langle p \rangle_t}{\partial t} = -m\Omega^2 \langle x \rangle_t - (2-r)C_{XP}\langle p \rangle_t$$

$$\frac{\partial \langle x^2 \rangle_t}{\partial t} = -2rC_{XP}\langle x^2 \rangle_t + \frac{2\langle xp \rangle_t}{m} + D_{PP}$$

$$\frac{\partial \langle xp \rangle_t}{\partial t} = -m\Omega^2 \langle x^2 \rangle_t - 2C_{XP}\langle XP \rangle_t + \frac{\langle p^2 \rangle_t}{m} - D_{XP}$$

$$\frac{\partial \langle p^2 \rangle_t}{\partial t} = -2m\Omega^2 \langle xp \rangle_t - (4-2r)C_{XP}\langle p^2 \rangle_t + D_{XX},$$

where the moments of the Wigner function are calculated as

$$\langle f(x, p) \rangle_t = \int_{-\infty}^{\infty} dx \int_{-\infty}^{\infty} dp \; f(x, p) W(x, p, t). \tag{5.12}$$

These moments correspond to symmetric ordering of the quantum mechanical operators x and p (Schleich 2001). In particular, note that the time-dependence is solely contained in the Wigner function, in agreement with the fact that we work in the Schrödinger picture.

The solutions for the first moments are

$$\langle x \rangle_t = e^{-C_{XP}t} \left[x_0 \cos(\beta_r t) + x_r^0 \sin(\beta_r t) \right], \tag{5.13}$$

$$\langle p \rangle_t = e^{-C_{XP}t} \left[p_0 \cos(\beta_r t) - p_r^0 \sin(\beta_r t) \right], \tag{5.14}$$

where

$$x_r^0 = \frac{mC_{XP}x_0(1-r) + p_0}{m\beta_r}, \tag{5.15}$$

$$p_r^0 = \frac{C_{XP}p_0(1-r) + mx_0\Omega^2}{\beta_r}, \tag{5.16}$$

with

$$x_0 \equiv \langle x \rangle_0, \quad p_0 \equiv \langle p \rangle_0, \tag{5.17}$$

and

$$\beta_r \equiv \sqrt{\Omega^2 - C_{XP}^2(r-1)^2}. \tag{5.18}$$

Similar solutions have been presented in the works of Kumar et al. (2009), Săndulescu and Scutaru (1987), Isar et al. (1994). Equation (5.11) may alternatively be written in terms of the kinetic momentum

$$\langle \tilde{p} \rangle_t = \langle p \rangle_t - mrC_{XP}\langle x \rangle_t. \tag{5.19}$$

It follows

$$\frac{\partial \langle x \rangle_t}{\partial t} = \frac{\langle \tilde{p} \rangle_t}{m}, \tag{5.20}$$

$$\frac{\partial \langle \tilde{p} \rangle_t}{\partial t} = -m \left[\Omega^2 - r(r-2)C_{XP}^2 \right] \langle x \rangle_t - 2C_{XP} \langle \tilde{p} \rangle_t,$$

or equivalently gathered in the compact form

$$\frac{\partial^2 \langle x \rangle_t}{\partial t^2} + 2C_{XP} \frac{\partial \langle x \rangle_t}{\partial t} + \left[\Omega^2 - r(r-2)C_{XP}^2 \right] \langle x \rangle_t = 0. \tag{5.21}$$

which, of course, can be derived directly from the equations Eq. (5.11). For both $r = 0$ and $r = 2$ one obtains a damped oscillator with the original frequency of the harmonic trap, Ω. For other values of r the frequency is renormalized, with the maximal renormalization corresponding to $r = 1$.

Equation (5.11) we see that r introduces apparent damping in the position, as already noted by Wiseman and Munro (1998). Because of this, in the following we will set $r = 0$. The extra term proportional to D_{PP}, not present in the starting Born-Markov master equation, appears only in the equation for $\langle x^2 \rangle$, without affecting the other equations, and in particular those for the first moments, so that it may be interpreted as a *position diffusion coefficient*.

We focus now on the stationary solution of Eq. (5.10). The latter may be found by means of the following Gaussian ansatz

$$W_{ST} = \zeta \exp \left[\frac{1}{2(\rho^2 - 1)} \left(\frac{x^2}{\sigma_x^2} + \frac{p^2}{\sigma_p^2} + \frac{2\rho x p}{\sigma_x \sigma_p} \right) \right], \tag{5.22}$$

which is normalized to one taking

$$\zeta \equiv \frac{1}{2\pi \sigma_x \sigma_p \sqrt{1 - \rho^2}}, \quad |\rho| \le 1, \tag{5.23}$$

with

$$\sigma_x = \sqrt{\langle x^2 \rangle}, \quad \sigma_p = \sqrt{\langle p^2 \rangle}, \quad \rho = -\frac{\langle xp \rangle}{\sigma_x \sigma_p}, \tag{5.24}$$

and, in the remainder of this Section, the variances are computed using the time-independent Gaussian Ansatz in Eq. (5.22) of Weedbrook et al. (2012). Inserting the Gaussian ansatz in Eq. (5.22) into Eq. (5.10) we find:

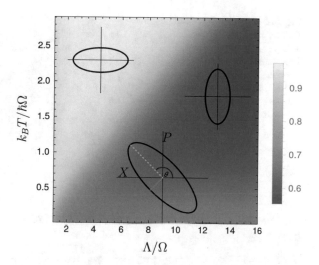

Fig. 5.1 Plot of the angle θ/π at $\gamma/\Omega = 0.8$. This angle is represented in the ellipse at the bottom of the picture. Here, the orange-solid (green-dashed) line represents the minor (major) axis of the Wigner function, i.e. that related to δ_l (δ_L). The axes X and P are those of the phase space

$$\sigma_x^2 = \frac{D_{XX} - 4mC_{XP}D_{XP} + m^2(4C_{XP}^2 + \Omega^2)D_{PP}}{4m^2C_{XP}\Omega^2} \tag{5.25}$$

$$\sigma_p^2 = \frac{D_{XX} + m^2\Omega^2 D_{PP}}{4C_{XP}}$$

$$\sigma_p\sigma_p\rho = mD_{PP}/2.$$

Again, we introduce the dimensionless variables

$$\delta_x = \sqrt{\frac{2m\Omega\sigma_x^2}{\hbar}}, \quad \delta_p = \sqrt{\frac{2\sigma_p^2}{m\Omega\hbar}}. \tag{5.26}$$

With this parametrization, the Heisenberg inequality $\sigma_x\sigma_p \geq \hbar/2$ reads $\delta_x\delta_p \geq 1$.

The Lindbladian character of Eq. (5.10) guarantees that the second moments will satisfy the Heisenberg relation at all times. We furthermore note that the term with coefficient D_{PP}, i.e. the extra term induced by the Lindblad form of the ME, leads to a correlation between the two canonical variables.

Geometrically, this correlation can be interpreted as a rotation of the stationary solution in the phase space, see the black sketches in Fig. 5.1. In the CL limit, the term with the coefficient D_{PP} is negligible, and the solution is an ellipse with its axes parallel to the canonical ones, reproducing the well-known results.

To analyze the properties of the stationary state in the phase space, we consider the variances of the major and minor axes of the Wigner function. These axes are defined as the eigenvectors of the covariance matrix

$$\text{cov}(X, P) = \begin{pmatrix} \delta_x^2 & -\rho\delta_x\delta_p \\ -\rho\delta_x\delta_p & \delta_p^2 \end{pmatrix}. \tag{5.27}$$

Fig. 5.2 Eccentricity of the Wigner function introduced in Eq. (5.22), at $\gamma/\Omega = 0.8$. The red dashed line represents the values of T and Λ yielding $\delta_l^2 = 1$, and we have genuine squeezing below it

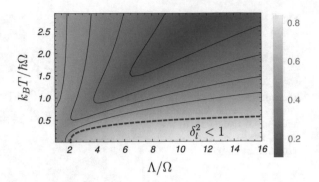

The smaller and larger eigenvalues of this matrix, δ_l and δ_L, are given respectively by:

$$\delta_{l,L}^2 = \frac{1}{2}\left(\delta_x^2 + \delta_p^2 \mp \sqrt{\left(\delta_x^2 - \delta_p^2\right)^2 + 4\delta_x^2\delta_p^2\rho^2}\right). \tag{5.28}$$

We now aim to quantify such a rotation, calculating the angle θ between the major axis of the Wigner function (i.e. the eigenvector corresponding to δ_L), and the x-axis of the phase space. In Fig. 5.1 we present the behavior of θ as function of T and Λ, at fixed γ. At high Λ the major axis aligns approximately with the p-axis of the phase space ($\theta = \pi/2$), while at low Λ, it is close to the x-axis ($\theta = \pi$), in agreement with the behavior of the Born-Markov master equation discussed by Massignan (2015), where $\langle xp \rangle$ was identically zero. On the other hand, at low temperatures the Wigner function associated to the stationary solution of the Lindblad equation may be significantly rotated with respect to the axes of the phase space.

In the previous chapters it has been shown that, in the low temperature regime, the position of the Brownian particle governed by the Born-Markov master equation experiences *genuine squeezing* along x in the Wigner function representation, i.e. $\delta_x < 1$. Similar squeezing effects are pointed out by Maniscalco et al. (2004), by studying the numerical solution of the exact master equation. In the case of the Lindblad equation, it was checked numerically that δ_x introduced in Eq. (5.25) is always bigger than one. However, the minor axis of the ellipse describing the Wigner function can display genuine squeezing. To quantify the degree of squeezing of the Wigner function, Fig. 5.2 shows the values of eccentricity defined as

$$\eta = \sqrt{1 - (\delta_l/\delta_L)^2}, \tag{5.29}$$

computed for different values of temperature T and UV-cut-off Λ. The eccentricity is larger at low temperatures. In particular, below the red dashed line, we find an area where $\delta_l < 1$, corresponding to genuine squeezing along the minor axis of the Wigner Function, while in the Caldeira-Leggett limit the eccentricity η approaches zero, and we obtain a Wigner function with circular symmetry. In Fig. 5.3 we present the minimal value of δ_l^2 obtained by choosing the appropriate (low) temperature. This

Fig. 5.3 Minimum value of δ_l^2 over all temperatures, as a function of the cut-off frequency, at several values of the damping constant

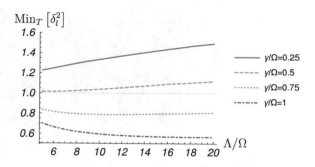

picture highlights the range of values of Λ and γ where genuine squeezing occurs. We find that the eccentricity is an increasing function of the damping constant, i.e. squeezing becomes more pronounced as γ grows. In particular, at least $\gamma/\Omega > 0.5$ is needed to obtain $\delta_l < 1$.

We look now into the cooling effect introduced in Chap. 3. Recalling Eq. (3.83), we thus define the system to be cooled if[1]

$$\chi = \frac{\delta_l \delta_L}{\coth\left(\frac{\hbar\Omega}{2k_B T}\right)} < 1, \qquad (5.30)$$

and heated otherwise. The degree of heating/cooling χ is shown in Fig. 5.4. In Fig. 5.5 we present the minimal value achieved by χ as the temperature is varied. We note that to obtain small values of χ one needs to choose large values of both Λ and γ.

There is a difference between the configuration of the cooling areas arising in the Lindblad dynamics studied here, and the ones produced by Eq. (3.36) studied in Chap. 3. In the latter, the cooling/heating boundary coincides with the line defined by $\delta_x = \delta_p$, and this condition does not depend on γ, while in the present Lindblad model, the location of the boundary varies with γ. However, the boundary calculated within the Lindblad approach converges to that predicted in the Born-Markov one in the $\gamma \to 0$ limit. Moreover, the Lindblad equation discussed here displays heating at very low temperatures.

In Figs. 5.3 and 5.5 we have not extended the range of values of the damping constant beyond $\gamma = 1$. In fact, the expressions for the coefficients of Eq. (5.3) have been obtained by comparing it with the Eq. (3.36). The latter is perturbative to second order in the strength of the coupling between the Brownian particle and the environment. The square of the coupling constant is proportional to the damping coefficient, so the validity of the perturbative expansion fails for γ large. In particular, in the case of QBM this perturbative expansion holds for $\gamma \lesssim \Omega$ (Breuer and Petruccione 2007; Haake et al. 1985).

---

[1]For the Gibbs-Boltzmann distribution we have $\langle X^2 \rangle_{GB} \langle P^2 \rangle_{GB} \sim \coth^2 (\hbar\Omega/2k_B T)$. So the denominator of Eq. (5.30) provides information regarding the area of the Gibbs-Boltzmann distribution.

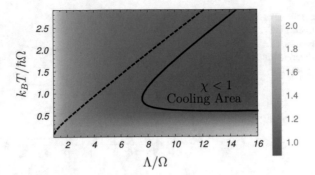

Fig. 5.4 Cooling parameter χ introduced in Eq. (5.30), plotted for $\gamma/\Omega = 0.8$. The system exhibits cooling to the right of the solid line, and heating to its left. For comparison, the dashed line represents the cooling/heating boundary obtained with the Born-Markov master equation (3.36), which is independent of γ

Fig. 5.5 Minimum value of the cooling parameter χ over all temperatures, as a function of the cut-off frequency, at several values of the damping constant

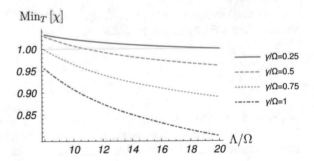

5.1.1.1   Low Temperature Regime

We consider here in detail the stationary state in the low temperature regime $k_B T < \hbar\Omega$. Such a study was impossible in Chap. 3 because solutions violated the Heisenberg principle there. Here, the Lindblad form of Eq. (5.3) ensures the positivity of the density matrix at all times, so no violations of the Heisenberg principle occur.

In the discussion above, we noticed that the time-dependent equations of motion of the Lindblad equation admit as an exact solution a Gaussian with non-zero correlations between the two canonical variables x and p. In the stationary state, in particular, one finds $\langle xp \rangle = -mD_{PP}/2 \neq 0$. This is a novelty in comparison with the stationary solution of Eq. (3.36), which shows no correlations between x and p. In the range of Λ explored in Fig. 5.1, the correlation between x and p becomes noticeable for $k_B T \lesssim 0.5\hbar\Omega$. So, an important feature of the stationary solution of the Lindblad equation at low temperature is that its major axis is rotated with respect to those of the phase space.

In Fig. 5.2 we analyze the eccentricity of the stationary state. We point out that as the temperature decreases, the distribution becomes increasingly more squeezed. In particular, at low temperature we find a region displaying genuine squeezing of the probability distribution in the direction of l. In Fig. 5.4 we also note the presence of a cooling area in the low temperature regime. Nevertheless, in the zero-temperature limit the stationary state shows again heating.

The zero-temperature limit of the Lindblad model deserves special attention, as the two limits $T \to 0$ and $\gamma \to 0$ do not commute. Taking first the zero-coupling and then the zero-temperature limit, one simply finds $\delta_x = \delta_p$ (in agreement with the general result for a free harmonic oscillator), but no further information on their specific value. If instead one takes first $T \to 0$ and then $\gamma \to 0$, one finds $\delta_x = \delta_p$ and the additional condition

$$\delta_x \delta_p = \delta_l \delta_L = \frac{5}{4} + \frac{\left[\log(\Lambda/\Omega)\right]^2}{\pi^2} > 1, \tag{5.31}$$

indicating that for the Lindblad model the Heisenberg inequality is not saturated in the limit when the particle becomes free. This is in contrast with the behavior of the non-Lindblad equation (3.36), for which, in this limit, we have $\delta_x \delta_p = 1$. Summarizing, the effect of D_{PP} is to introduce extra heating at low temperatures and couplings, manifested by a small constant, and a weak logarithmic dependence on the ultraviolet cut-off Λ.

5.2 Quadratic Case

5.2.1 The Hamiltonian and the Lindblad Equation

In this section we consider the quadratic QBM, whose coupling is still linear in the positions of the oscillators of the bath, but is quadratic in the position of the Brownian particle

$$H_I = \sum_k \frac{g_k}{a} X_k x^2. \tag{5.32}$$

Here a is a characteristic length related to the motion of the Brownian particle and we set it to be $a = \sqrt{\hbar/m\Omega}$. The interaction term in Eq. (5.32) describes an interaction of the particle with an inhomogeneous environment, giving rise to position-dependent damping and diffusion.

The dynamics induced by the interaction term in Eq. (5.32) has already been discussed in detail in Chap. 4. There, the master equation for the Brownian particle has been derived in the Born-Markov approximations, for a Lorentz-Drude spectral density. Nevertheless, this master equation is not in a Lindblad form, nor is exact. Accordingly, the stationary solution is not defined for some values of the model parameters because of violations of the Heisenberg uncertainty principle at low temperatures.

In this Section, we aim to find a Lindblad equation as similar as possible to that derived in Chap. 4. Just like in the case of linear QBM, we expect it to differ from the Born-Markov by some extra terms. To achieve this goal, we consider a single Lindblad operator

$$A_1 = \mu x^2 + \nu\{x, p\} + \epsilon p^2, \tag{5.33}$$

where μ, ν and ϵ are non-zero complex numbers. Substituting it into Eq. (5.1) we obtain:

$$\frac{\partial \rho}{\partial t} = -\frac{i}{\hbar}[H_S + \Delta H_2, \rho] \tag{5.34}$$

$$-\frac{D_{\mu\nu}}{\hbar^2}\left[x^2, [\{x, p\}, \rho]\right] - \frac{D_{\mu\epsilon}}{\hbar^2}\left[x^2, [p^2, \rho]\right] - \frac{D_{\epsilon\nu}}{\hbar^2}\left[p^2, [\{x, p\}, \rho]\right]$$

$$-i\frac{C_{\mu\nu}}{\hbar}\left[x^2, \{\{x, p\}, \rho\}\right] - i\frac{C_{\mu\epsilon}}{\hbar}\left[x^2, \{p^2, \rho\}\right] - i\frac{C_{\epsilon\nu}}{\hbar}\left[p^2, \{\{x, p\}, \rho\}\right]$$

$$-\frac{D_\mu}{2\hbar^2}\left[x^2, [x^2, \rho]\right] - \frac{D_\nu}{2\hbar^2}\left[\{x, p\}, [\{x, p\}, \rho]\right] - \frac{D_\epsilon}{2\hbar^2}\left[p^2, [p^2, \rho]\right]$$

where

$$\frac{D_\mu}{\hbar^2} \equiv |\mu^2|, \quad \frac{D_{\mu\nu}}{\hbar^2} \equiv \mathrm{Re}(\mu^*\nu), \quad \frac{C_{\mu\nu}}{\hbar} \equiv \mathrm{Im}(\mu^*\nu), \tag{5.35}$$

and similarly for the other combinations of indices. We could have obtained the same result by means of three Lindblad operators (rather than a single one), each proportional to one of the terms appearing on the right-hand side of Eq. (5.33). Similarly to the linear case, there is a term which appears in the unitary part of the master equation

$$H_2 = 2D_{\mu\nu}x^2 - 2D_{\epsilon\nu}p^2 + 2D_{\mu\epsilon}\{x, p\} \tag{5.36}$$

$$-\frac{1}{2}C_{\mu\nu}\{\{x, p\}, x^2\} - \frac{1}{2}C_{\mu\epsilon}\{p^2, x^2\}$$

$$+\frac{1}{2}C_{\epsilon\nu}\{\{x, p\}, p^2\}.$$

We eliminate it by introducing appropriate counter terms in the Hamiltonian.

Equation (5.34) is in a Lindblad form. Proceeding as in Sect. 5.1, equating the coefficients on the right hand side of Eq. (5.34) to the corresponding ones in the Born-Markov master equation for quadratic QBM derived in Chap. 3, we obtain:

$$D_{\mu\epsilon} = \frac{D_{pp}}{m\Omega}, \quad D_{\mu\nu} = D_{xp}, \tag{5.37}$$

$$C_{\mu\epsilon} = \frac{C_{pp}}{\hbar m\Omega}, \quad C_{\mu\nu} = \frac{C_{xp}}{\hbar},$$

and $D_\mu = 2m\Omega D_{xx}$. The remaining coefficients are then uniquely determined as

$$D_{\epsilon\nu} = \frac{1}{D_\mu} \left[D_{\mu\nu} D_{\mu\epsilon} + \hbar^2 C_{\mu\nu} C_{\mu\epsilon} \right], \tag{5.38}$$

$$C_{\epsilon\nu} = \frac{1}{D_\mu} \left[C_{\mu\nu} D_{\mu\epsilon} - D_{\mu\nu} C_{\mu\epsilon} \right],$$

$$D_\epsilon = \frac{1}{D_\mu} \left[D_{\mu\epsilon}^2 + \left(\hbar C_{\mu\epsilon} \right)^2 \right],$$

$$D_\nu = \frac{1}{D_\mu} \left[D_{\mu\nu}^2 + \left(\hbar C_{\mu\nu} \right)^2 \right].$$

It is easy to check that in the limit $k_B T \gg \hbar\Lambda \gg \hbar\Omega$, the coefficients of all extra terms vanish, and Eq. (5.34) recovers the structure of Eq. (4.5).

5.2.2 Stationary State of the Quadratic Quantum Brownian Motion

We turn now to the study of the stationary state of the Brownian particle in the case of quadratic coupling. To this end we express the Lindblad master in Eq. (5.34) in terms of the Wigner function W, and obtain an equation of the form $\dot{W} = \mathcal{L}W$, with:

$$\mathcal{L} = -\frac{\partial_x p}{m} + m\Omega^2 \partial_p x + 2D_\mu \partial_p^2 X^2 + 2D_\nu \left(\partial_p p - \partial_x x \right)^2 + 2D_\epsilon \partial_x^2 p^2 \tag{5.39}$$

$$+ 4D_{\mu\nu}(\partial_p^2 xp - \partial_p \partial_x x^2 + \partial_p x) - 4D_{\epsilon\nu} p\partial_x \left(\partial_p p - \partial_x x \right)$$

$$+ 8C_{\mu\nu} \left[\partial_p px^2 + \frac{\hbar^2}{4} \partial_p^2 (\partial_x x - 1) \right] + C_{\mu\epsilon} \left[4\partial_p xp^2 - \hbar^2 \partial_p \partial_x^2 x + 2\hbar^2 \partial_p \partial_x \right]$$

$$- 2C_{\epsilon\nu} p\partial_x \left(4xp + \hbar^2 \partial_p \partial_x \right) - 4D_{\mu\epsilon}(\partial_x x - 1)\partial_p p.$$

We now find the stationary solution of the above equation. In this case the Gaussian ansatz in Eq. (5.22) may at best provide an approximate solution, in contrast with the case of the linear QBM, since the system of equations for the second moments is not closed. We approximate higher-order moments by their Wick expressions in terms of second moments (which would be exact in a Gaussian case), obtaining the following closed, non-linear system of equations in the variables δ_x, δ_p and ρ

$$\frac{1}{2} \frac{\partial \delta_x^2}{\partial t} = 4m\hbar\Omega C_{\epsilon\nu}[1 + \delta_x^2 \delta_p^2 (1 + 2\rho^2)] + 2m^2\Omega^2 D_\epsilon \delta_p^2$$

$$+ 4D_\nu \delta_x^2 - \Omega\delta_x\delta_p\rho, \tag{5.40}$$

$$\frac{1}{2}\frac{\partial \delta_p^2}{\partial t} = \frac{2D_\mu}{m^2\Omega^2}\delta_x^2 - \frac{4\hbar}{m\Omega}C_{\mu\epsilon} + 6\hbar C_{\mu\epsilon}\delta_x\delta_p^3\rho + \Omega\delta_x\delta_p\rho$$
$$+ 4\delta_p^2\left[D_\nu - D_{\mu\epsilon} - \frac{\hbar C_{\mu\nu}}{m\Omega}\left(1 + 2\rho^2\right)\delta_x^2\right], \tag{5.41}$$

and

$$-\frac{1}{2}\frac{\partial(\delta_x\delta_p\rho)}{\partial t} = 4\hbar C_{\mu\epsilon} + \Omega\delta_p^2 - 8m\Omega D_{\epsilon\nu}\delta_p^2 + \frac{12\hbar}{m\Omega}C_{\mu\nu}\delta_p\delta_x^3\rho$$
$$+ \left(8D_{\mu\epsilon} - 12m\hbar\Omega C_{\epsilon\nu}\delta_p^2\right)\delta_x\delta_p\rho - \left[\Omega + 8\frac{D_{\mu\nu}}{m\Omega} + 2\hbar\left(1 + 2\rho^2\right)C_{\mu\epsilon}\delta_p^2\right]\delta_x^2.$$
$$\tag{5.42}$$

This system of equations could admit more than one stationary solution, so we have to study the proper one. We choose the solution that coincides with that obtained with the non-Lindblad dynamics in the Caldeira-Leggett limit, since in this limit the coefficients of the extra terms of Eq. (5.34) vanish. Chapter 4 the stationary state in the case of the non-Lindblad dynamics has been studied in detail, and the variances have been calculated analytically.

Similarly to the linear QBM studied in the previous section, we characterize the stationary state in terms of the variances of the Wigner function, and define the eccentricity, the cooling parameter, and the angle between the major axis and the position axis of the phase space as before. These quantities are shown in Figs. 5.6, 5.7, and 5.8, as functions of Λ and T, when $\gamma/\Omega = 0.1$. In Fig. 5.6 we point out that the eccentricity tends to zero in the Caldeira-Leggett limit, while it increases away from it. This behavior is similar to that found for the linear QBM. We found that for $\gamma/\Omega \le 0.1$ the Brownian particle experiences neither cooling nor genuine squeezing.

In contrast to the linear case, we do not find a noticeable rotation at low temperature in the quadratic one. We would expect to observe this at larger values of γ, as in the case of linear coupling. However, for larger values of the damping constant the many stationary solutions of the system of Eqs. (5.40–5.42) cross, and therefore it is not

Fig. 5.6 Eccentricity η of the Wigner function at $\gamma/\Omega = 0.1$, for quadratic coupling

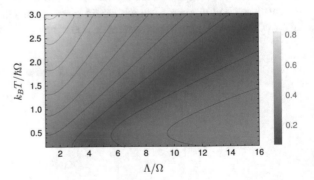

Fig. 5.7 Cooling parameter χ for quadratic coupling, at $\gamma/\Omega = 0.1$

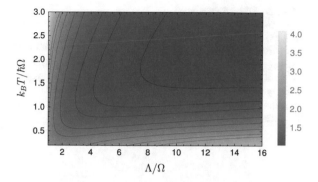

Fig. 5.8 Angle θ/π between the major axis of the Wigner function, and the X axis of the phase space at $\gamma/\Omega = 0.1$, for quadratic coupling

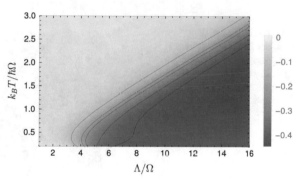

straightforward to determine the stationary solution of (5.39) that coincides with the one obtained in the Caldeira-Leggett limit. Moreover, for larger values of γ the Gaussian ansatz given in Eq. (5.22) may fail to approximate any stationary states. To show this point, in Fig. 5.9 we plotted the time dependence of δ_x^2 for several values of γ, at fixed values of T and Λ. Above a certain value of γ, the position variance does not converge to a stationary value. This suggests that in these cases the Gaussian solution of Eq. (5.39) is not stationary. Figure 5.9 is plotted for the initial conditions $\delta_x^2 = \delta_p^2 = 1$, corresponding to the case when the harmonic oscillator is in its ground state. The choice of the initial conditions is not crucial, as we observe a very similar behavior with quite different initial conditions.

We conclude this Section pointing out that, although in Eqs. (5.40–5.42) we performed the Gaussian approximation at the level of the equations for the moments, it is possible to obtain exactly the same result applying the approximation directly on the original equation in Eq. (5.34), or on that Lindblad equation expressed in terms of the Wigner function, Eq. (5.39). In Appendix B we show, by a very general analytical demonstration, that the Gaussian approximation applied to the original Lindblad equation yields again a master equation of the Lindblad form, guaranteeing therefore that the approximated solutions will preserve the Heisenberg principle at all times. We provide further numerical evidence of this fact in Fig. 5.10, where we

Fig. 5.9 Time dependence of δ_x^2 for several values of γ, at $\Lambda/\Omega = 16$ and $k_B T/\hbar\Omega = 4$. The thin solid lines represent the stationary value of δ_x^2 in the state, namely the stationary solution of Eqs. (5.40–5.42) for such a quantity

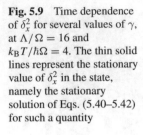

Fig. 5.10 Plot of the product $\delta_x \delta_p$ at $\gamma/\Omega = 0.1$, for quadratic coupling. This quantity is always larger than 1, in accordance with the Heisenberg principle

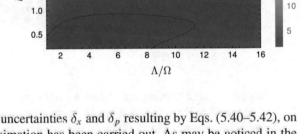

plot the product of the two uncertainties δ_x and δ_p resulting by Eqs. (5.40–5.42), on which the Gaussian approximation has been carried out. As may be noticed in the figure, the approximated equations do not produce any violation of the Heisenberg principle.

Chapter 6
Heisenberg Equations Approach

In the previous chapter we have proposed a Lindblad model for QBM, exploring both the cases of linear and non-linear coupling. Lindblad equations differ from the original Born-Markov ones just for a few extra-terms (only one term in the linear case), curing the forbidden area detected in Chaps. 3 and 4. In this way, it is possible to evaluate the correlation functions of both position and momentum at each temperature and for each value of the bath-system coupling strength. The problem of the Lindblad approach to QBM is that the corresponding equations cannot be obtained microscopically by a Hamiltonian model. For instance, in the context of polaron physics, where will apply the model in the next chapter, the physical Hamiltonian of the system does not lead to a master equation in the Lindblad form.

We go through another approach to investigate the dynamics of QBM model: Heisenberg equations formalism. We derive equations for the observables of the quantum Brownian particle, as well as those for the bath's operators, describing the behavior of the system in the Heisenberg picture. It is possible to note that they may be opportunely combined in order to obtain an equation ruling the temporal evolution of the particle position, where the influence of the bath appears in the form of noise and damping: it is a quantum stochastic equation. In particular, considering an ohmic Lorentz-Drude spectral density in the infinite cut-off limit, such an equation takes the form of the Langevin one in Eq. (2.18), for the classical Brownian motion: it shows a white noise and a local in time damping. The main goal of the chapter is to present in detail the procedure to solve this quantum Langevin equation, aiming to calculate the position variance. Here, we distinguish two cases: the situation in which the particle is in a harmonic potential and that where it is untrapped. In the latter, the particle runs away from the initial position. Such a behavior may be described by means of the mean square displacement discussed in Chap. 2, providing a signature of Brownian motion. We recover the traditional diffusive behavior, constituting here a consequence of the local in time form of the quantum Langevin equation, i.e. of the absence of memory effect in the dynamical behavior of the Brownian particle. If the particle is trapped in a harmonic potential, instead, it approaches an equilibrium state localized in average in the middle of the trap. We evaluate position and momentum variances related to such a long-time state, proving that the Heisenberg principle is always fulfilled and a proper analysis of the zero-temperature limit is finally possible.

This topic belongs to standard textbook material (Weiss 2008; Breuer and Petruccione 2007), and has been recently considered by Boyanovsky and Jasnow (2017). We shall follow the revisited treatment published by Lampo et al. (2017), discussing in detail the techniques we use. In the end of the chapter, we will extend the Heisenberg equations formalism to non-linear QBM, proceeding as in the work of Barik and Ray (2005). In this situation, to provide an analytical expression for the position variance is not an easy task so we limit to present the form of the equation without to solve it.

6.1 Derivation of the Heisenberg Equations in the Linear Case

In this section we consider the QBM Hamiltonian in the linear case, i.e. that with interaction term in Eq. (3.4). Such a model leads to the following equations of motion for the central particle and the environmental oscillators

$$\dot{x}(t) = \frac{i}{\hbar}[H, x(t)] = p(t)/m, \tag{6.1}$$

$$\dot{x}_n(t) = \frac{i}{\hbar}[H, x_n(t)] = p_n(t)/m_n, \tag{6.2}$$

$$\dot{p}(t) = \frac{i}{\hbar}[H, p(t)] = -H_c'(x(t)) + \sum_n \kappa_n x_n(t), \tag{6.3}$$

$$\dot{p}_n(t) = \frac{i}{\hbar}[H, p_n(t)] = -m_n \omega_n^2 x_n(t) + \kappa_n x(t), \tag{6.4}$$

with

$$H_c(x) = V_c(x) + U(x), \tag{6.5}$$

where V_c is the counter-term introduced in Eq. (3.5) and $U(x)$ is the bare impurity potential. For sake of clearness we underline that, differently by the previous chapters, the operators of the system are no longer in the Schrödinger picture, but in the Heisenberg one. It is easy to find that the equation for the coordinate of the Brownian particles is

$$m\ddot{x}(t) + H_c'(x(t)) - \sum_n \kappa_n x_n(t) = 0, \tag{6.6}$$

while the equations for the coordinates of the bath oscillators take the form

$$m\ddot{x}_n(t) + m_n\omega_n^2 x_n^2(t) - \kappa_n x(t) = 0. \tag{6.7}$$

The last equation shows that the nth bath oscillator is driven by the force $\kappa_n x(t)$ which depends linearly on the coordinate of the Brownian particle. In order to get a closed equation of motion for $x(t)$ the first step is to solve Eq. (6.7) in terms of $x(t)$ and of the initial conditions for the bath modes. The solution of this equation is the sum of that of the related homogeneous equation plus the particular one, that may be expressed as convolution product of Green function and the particle position. This calculation is very well known (see for instance Appendix A, where the procedure is presented in detail for the concrete case of the polaron) and yields

$$x_n(t) = \sqrt{\frac{\hbar}{2m_n\omega_n}}\left(e^{-i\omega_n t}b_n + e^{i\omega_n t}b_n^\dagger\right) + \frac{\kappa_n}{m_n\omega_n}\int_0^t ds \sin\left[\omega_n(t-s)\right]x(s). \tag{6.8}$$

where we have introduced again the bath creation and annihilation operators

$$x_n(t) = \sqrt{\frac{\hbar}{2m_n\omega_n}}\left(b_n + b_n^\dagger\right), \quad p_n(t) = -i\sqrt{\frac{m_n\omega_n\hbar}{2}}\left(b_n - b_n^\dagger\right). \tag{6.9}$$

Replacing Eq. (6.8) in Eq. (6.6) we obtain

$$m\ddot{x}(t) + H_c'(x(t)) - \sum_n \frac{\kappa_n^2}{m_n\omega_n}\int_0^t ds \sin\left[\omega_n(t-s)\right]x(s) = B(t), \tag{6.10}$$

where we recall that the operator $B(t)$ appearing here on the right-hand side is

$$B(t) = \sum_n \kappa_n \sqrt{\frac{\hbar}{2m_n\omega_n}}\left(e^{-i\omega_n t}b_n + e^{i\omega_n t}b_n^\dagger\right), \tag{6.11}$$

representing the temporal evolution of the Schrödinger operator $B = \sum_n \kappa_n x_n(0)$. With the help of the dissipation kernel (3.30) in Eq. (6.10) may be cast in the form

$$\ddot{x}(t) + \frac{1}{m}H_c'(x(t)) - \frac{1}{\hbar m}\int_0^t ds\,\eta(t-s)x(s) = \frac{1}{m}B(t). \tag{6.12}$$

In the theory of QBM it is useful to express the dissipation kernel in terms of another quantity which is known as the damping kernel

$$\Gamma(t-s) = \frac{1}{m}\int_0^\infty d\omega\,J(\omega)\cos[\omega(t-s)], \tag{6.13}$$

fulfilling

$$\frac{\partial \Gamma}{\partial t} = -\frac{1}{\hbar m}\eta(t - s), \tag{6.14}$$

and

$$\Gamma(0) = \frac{1}{m}\int_0^\infty d\omega \frac{J(\omega)}{\omega} = \sum_n \frac{\kappa_n^2}{mm_n\omega_n^2}, \tag{6.15}$$

where $J(\omega)$ is the spectral density introduced in Chap. 3 in Eq. (3.31). With the help of the damping kernel we may write the dissipative term in Eq. (6.12) as follows

$$-\frac{1}{\hbar m}\int_0^t ds\,\eta(t - s)x(s) = \int_0^t ds \frac{\partial}{\partial t}\Gamma(t - s)x(s) = \tag{6.16}$$

$$= \frac{\partial}{\partial t}\int_0^t ds\,\Gamma(t - s)x(s) - \Gamma(0)x(t). \tag{6.17}$$

In view of Eq. (6.14) the last term $-\Gamma(0)x(t)$ is seen to cancel the contribution from the counter-term contained in the potential H_c. Thus we finally arrive at the following exact Heisenberg equation of motion,

$$\ddot{x}(t) + \frac{1}{m}U'(x(t)) + \frac{\partial}{\partial t}\int_0^t ds\,\Gamma(t - s)x(s) = \frac{1}{m}B(t). \tag{6.18}$$

Equation (6.18) is the equation of motion for the coordinate of the Brownian particle. It may be viewed as the quantum analogue of a classical stochastic differential equation, involving a damping kernel $\Gamma(t - s)$ and a stochastic force $B(t)$, whose statistical properties depend on the initial distribution at $t = 0$. If we consider now the ohmic spectral density with a Lorentz-Drude cut-off (3.41) we have

$$\Gamma(t - s) = \gamma\Lambda\exp(-\Lambda t). \tag{6.19}$$

In the limit $\Lambda \to \infty$ it takes the following form

$$\Gamma(t - s) = 2\gamma\delta(t - s). \tag{6.20}$$

We recognize so the physical meaning of the constants introduced in Eq. 3.41. The damping kernel in Eq. (6.19) induces a dependence of the particle dynamics on its past history, decaying according a timescale given by $1/\Lambda$. We may say thus that it is the characteristic time ruling the lost of memory effects, i.e. over which the behavior of the quantum particle may be considered Markovian. The constant γ is the responsible for the damping process, i.e. the lost of energy, we interpret therefore $1/\gamma$ as the relaxation timescale associated to dissipation.

Replacing Eq. (6.20) into (6.18), the latter writes as

$$\ddot{x}(t) + \frac{1}{m}U'(x(t)) + \gamma\dot{x}(t) = \frac{1}{m}B(t). \qquad (6.21)$$

We recover so the equation derived by Langevin in the classical context, presented in Eq. (2.18). This result justifies the name of the Hamiltonian model we are dealing with: quantum Brownian motion. In fact, when in Chap. 3 we proposed a Hamiltonian to study the quantum version of the phenomenon we looked for an operator yielding to dynamical equations manifesting the same form of the phenomenological ones discussed in Chap. 2. The fact that the operator in Eq. (3.1) with the bilinear interaction (3.4) leads to a Langevin-type equation endorses the appropriateness of such a Hamiltonian to study the Brownian motion in the quantum regime. Of course, to justify the Hamiltonian model we use, it is not necessary to switch to the quantum regime. One could indeed work in a classical framework with canonical equations, obtaining a functional equation with the form in Eq. (2.18). This was actually the historical procedure: to write a functional classical Hamiltonian such that Brownian motion equations arise, and then, switching from canonical variables to operator ones, one gets Hamiltonian (3.1).

Equation (6.21) is local in time, so it is free of memory effects, i.e. it corresponds to a pure Markovian dynamics. This is not only consequence of the ohmic character of the spectral density, because one has also to consider the large cut-off limit. From the physical point of view such a limit corresponds to focus on the long-time behavior of the particle, thus we may state that in the long-time the dynamics of the quantum particle does not carry memory effects.

However, in the general case of a non-ohmic spectral density Eq. (6.18) presents a dependence on the past-history. This is for instance the case of the polaron dynamics studied by Lampo et al. (2017), where the particular form of the bath-impurity coupling leads to a spectral density proportional to the cubic power of the frequency

$$J \sim \omega^3. \qquad (6.22)$$

Such a situation corresponds to an impurity embedded in a homogeneous condensate in one dimension. When the gas is not so, i.e. it is confined in a trapping potential making its density profile space dependent (inhomogeneous), it has been proven (see Lampo et al. 2018) that the spectral density writes

$$J \sim \omega^4. \qquad (6.23)$$

This means that the inhomogeneous character of the medium in this case produces an increase of the memory effects amount in the polaron dynamics.

6.2 Solution of the Heisenberg Equations in the Linear Case

The purpose of the present section is to solve Eq. (6.21). For this goal, we distinguish between the cases in which the Brownian particle is trapped in a harmonic potential and that where it is free of any trap.

6.2.1 Brownian Particle Trapped in a Harmonic Potential

We start by considering the situation where the particle is trapped in a quadratic potential

$$U(x) = \frac{1}{2}m\Omega^2 x^2. \tag{6.24}$$

Accordingly Eq. (6.21) takes the following form

$$\ddot{x}(t) + \Omega^2 x(t) + \frac{\partial}{\partial t} \int_0^t ds \Gamma(t-s)x(s) = \frac{1}{m}B(t). \tag{6.25}$$

This equation may be treated switching to the Laplace transform domain. Recalling the transformation properties of the derivatives, the solution follows directly

$$x(t) = G_1(t)x(0) + G_2(t)\dot{x}(0) + \frac{1}{m}\int_0^t G_2(t-s)B(s), \tag{6.26}$$

where G_1 and G_2 are defined by means of their Laplace transforms:

$$\mathcal{L}_z[G_1(t)] = \frac{z}{z^2 + \Omega^2 + z\mathcal{L}_z[\Gamma(t)]}, \quad \mathcal{L}_z[G_2(t)] = \frac{1}{z^2 + \Omega^2 + z\mathcal{L}_z[\Gamma(t)]}, \tag{6.27}$$

fulfilling the initial conditions

$$G_1(0) = 1, \quad G_2(0) = 0, \tag{6.28}$$

$$\dot{G}_1(0) = 0, \quad \dot{G}_2(0) = 1. \tag{6.29}$$

The Laplace transforms in Eq. (6.27) maybe easily inverted for an ohmic spectral density, namely for the damping kernel in Eq. (6.20) which Laplace transform is given by the constant γ. We obtain

$$G_1(t) = e^{-\gamma t} \left[\tilde{\Omega} \sinh\left(\gamma \tilde{\Omega} t\right) + \cosh\left(\gamma \tilde{\Omega} t\right) \right], \tag{6.30}$$

$$G_2(t) = \frac{e^{-\gamma t}}{\gamma \tilde{\Omega}} \sinh\left(\gamma \tilde{\Omega} t\right), \tag{6.31}$$

where we introduced the dimensionless parameter

$$\tilde{\Omega} = \sqrt{1 - (\Omega/\gamma)^2}, \tag{6.32}$$

ruling the qualitative behavior of the motion. In fact, if $\Omega > \gamma$, the quantity in Eq. (6.32) becomes imaginary and the particle performs oscillations. This does not occur if $\Omega \leq \gamma$. However, the presence of the exponential $e^{-\gamma t}$ in both expressions implies that, in any case, the contribution of the initial conditions exponentially vanishes in a range equals to the relaxation timescale: in general, oscillating or not, we have an exponentially damped motion.

We aim to characterize the long-time behavior of the particle by means of the position variance. Taking into account the fact that both G_1 and G_2 functions vanish at long times we infer by Eq. (6.26)

$$\langle x^2(t) \rangle = \int_0^t ds \int_0^t d\sigma G_2(t-s) G_2(t-\sigma) \frac{\nu(s-\sigma)}{m_I^2 \hbar^{-1}}, \tag{6.33}$$

where we used

$$\langle \{B(s), B(\sigma)\} \rangle = \nu(s-\sigma) = \frac{m\gamma\hbar}{\pi} \int_0^\Lambda d\omega\omega \coth\left(\hbar\omega/2k_B T\right) \cos\left[\omega(t-t')\right], \tag{6.34}$$

where ν is the noise kernel introduced in Eq. (3.29). Replacing the expression for the noise kernel one gets

$$\langle x^2(t) \rangle = \frac{\hbar}{m_I^2} \int_0^\Lambda J(\omega) \coth\left(\hbar\omega/2k_B T\right) d\omega$$

$$\times \int_0^t ds \int_0^t d\sigma G_2(t-s) G_2(t-\sigma) \cos[\omega(s-\sigma)]$$

$$\equiv \frac{\hbar}{m_I^2} \int_0^\Lambda J(\omega) \coth\left(\hbar\omega/2k_B T\right) \phi(t,\omega) d\omega. \tag{6.35}$$

Equation (6.35) turns into

$$\phi(t, \omega) = \frac{1}{2} \int_0^t ds \int_0^t d\sigma G_2(t-s) G_2(t-\sigma)$$
$$\times \left[e^{i\omega s} e^{-i\omega\sigma} + c.c. \right]$$
$$= \frac{1}{2} \int_0^t d\tilde{s} e^{-i\omega\tilde{s}} G_2(\tilde{s})$$
$$\times \int_0^t d\tilde{\sigma} e^{i\omega\tilde{\sigma}} G_2(\tilde{\sigma}) + c.c., \tag{6.36}$$

where we introduced

$$\tilde{s} = t - s, \quad \tilde{\sigma} = t - \sigma. \tag{6.37}$$

We are interested in the long-time limit, $t \to \infty$. In this limit, one gets

$$\phi(t, \omega) = \mathcal{L}_{-i\omega} [G_2(t)] \mathcal{L}_{+i\omega} [G_2(t)]. \tag{6.38}$$

Replacing the expression for $\mathcal{L}_z[G_2(t)]$ in Eq. (6.27) into Eq. (6.38), we obtain the final expression for the position variance:

$$\langle x^2 \rangle = \frac{\hbar}{2\pi} \int_{-\Lambda}^{+\Lambda} d\omega \coth (\hbar\omega/2k_B T) \tilde{\chi}''(\omega), \tag{6.39}$$

where

$$\tilde{\chi}''(\omega) = \frac{1}{m_{\mathrm{I}}} \frac{\zeta(\omega)\omega}{[\omega\zeta(\omega)]^2 + [\Omega^2 - \omega^2 + \omega\theta(\omega)]^2}, \tag{6.40}$$

and

$$\zeta(\omega) = \mathrm{Re}\{\mathcal{L}_{\tilde{z}} [\Gamma(t)]\} = \gamma + o \left(\frac{\omega}{\Lambda} \right)^2, \tag{6.41}$$

$$\theta(\omega) = \mathrm{Im}\{\mathcal{L}_{\tilde{z}} [\Gamma(t)]\} = 0 + o \left(\frac{\omega}{\Lambda} \right)^2. \tag{6.42}$$

The expression in Eq. (6.39), endowed by Eqs. (6.40) and (6.41), completely determines the position variance for an impurity trapped in a harmonic potential. In the same manner one may obtain an expression for the momentum variance:

$$\langle p^2 \rangle = m^2 \frac{\hbar}{2\pi} \int_{-\Lambda}^{+\Lambda} d\omega \omega^2 \coth (\hbar\omega/2k_B T) \tilde{\chi}''(\omega). \tag{6.43}$$

We emphasize that these expressions has been obtained just by considering the long-time limit of the solution of the Heisenberg equations in Eq. (6.26). It is possible to note, however, that it corresponds to that achieved in the context of the linear response theory by means of the fluctuation-dissipation theorem Breuer and Petruccione (2007). Indeed, $\tilde{\chi}''$ can be seen as the imaginary part of the Fourier transform of the linear response to an external force applied to the system, at the equilibrium.

In conclusion, in presence of a harmonic trap the impurity approaches the equilibrium in the long-time limit. We describe such a state through position and momentum variance (a similar expression to Eq. (6.39) is also found for the momentum). In particular we look to

$$\delta_x = \sqrt{\frac{2m_I \Omega \langle x^2 \rangle}{\hbar}}, \quad \delta_p = \sqrt{\frac{2 \langle p^2 \rangle}{m_I \hbar \Omega}}, \tag{6.44}$$

which represents the position and momentum variances regularized in order to be dimensionless. In these units the Heisenberg principle reads as $\delta_x \delta_p \geq 1$, so the Heisenberg threshold is set to be equal to one. These quantities do not depend on time, because they refer to an equilibrium stationary state. We study the dependence of δ_x on the system parameters, such as temperature and interaction strength, that can be tuned in experiments.

These quantities are plotted in Figs. 6.1 and 6.2. Both variances show an agreement with the equipartition theorem as the temperature grows. At low temperature, instead, the position variance becomes smaller than the value associated to the Heisenberg threshold, i.e. it exhibits genuine position squeezing. Such an effect, corresponding to high spatial localization of the particle is enhanced increasing the value of the system-bath coupling. In this approach, differently by those adopted in the previous chapters, the low-temperature limit does not alter the Heisenberg principle. This is clear in Fig. 6.3, where we plotted the product of position and momentum variance. The fact that

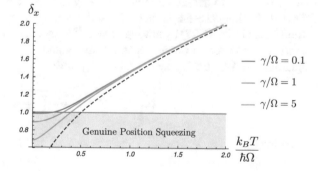

Fig. 6.1 Position variance for the ohmic spectral density in Eq. 3.41 in the large cut-off limit. The red dashed line represents the behavior predicted by the Equipartition theorem

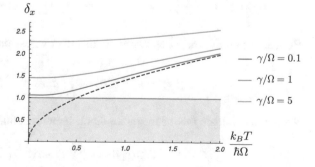

Fig. 6.2 Momentum variance for the ohmic spectral density in Eq. 3.41 in the large cut-off limit The red dashed line represents the behavior predicted by the Equipartition theorem

Fig. 6.3 Product of position and momentum variance for the ohmic spectral density in Eq. 3.41 in the large cut-off limit. The red dashed line represents the behavior predicted by the Equipartition theorem

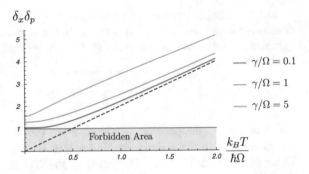

Heisenberg principle is preserved at any temperature is one of the most important result of this chapter and shows the vantage of the Heisenberg equations formalism to explore concrete systems, for instance the polaron, as we shown in the work of Lampo et al. (2017) Lampo et al. (2018). We finally mention that the analytical expressions we found for the correlation functions of position and momentum could be derived also by means of the fluctuation-dissipation theorem.

6.2.2 Untrapped Brownian Particle

We move now to the situation where the central particle is untrapped, i.e. we have $\Omega = 0$. In this case the kernels G_1 and G_2 may found by inverting the Laplace transforms in Eq. (6.27) by putting $\Omega = 0$. The calculation can be done immediately: it follows that for an ohmic spectral density and for a very large large cut-off the quantities in Eq. (6.27) take the following form

$$G_1(t) = 1, \quad G_2(t) = \frac{1}{\gamma}\left(1 - e^{-\gamma t}\right), \tag{6.45}$$

yielding to

$$x(t) = x(0) + \frac{1}{\gamma}\left(1 - e^{-\gamma t}\right)\dot{x}(0) + \int_0^t ds \frac{1}{\gamma m}\left[1 - e^{-\gamma(t-s)}\right] B(s). \tag{6.46}$$

Remembering Eq. (6.1) one easily obtains the expression for the momentum:

$$p(t) = e^{-\gamma t} p(0) + \int_0^t ds\, e^{-\gamma(t-s)} B(s). \tag{6.47}$$

We aim to evaluate the mean square displacement, defined as

$$\text{MSD}(t, t') = \langle\left[x(t) - x(t')\right]^2\rangle. \tag{6.48}$$

We replace in this expression that in Eq. (6.46) obtaining

$$\text{MSD}(t, t') = \frac{1}{m\gamma^2} \left(e^{-\gamma t} - e^{-\gamma t'} \right) \langle p^2(0) \rangle \tag{6.49}$$

$$+ \frac{\gamma\hbar}{m\pi} \int_0^\Lambda \frac{\omega}{\omega^2 + \gamma^2} \coth\left(\frac{\hbar\omega}{2k_B T}\right) \left| \frac{e^{i\omega t} - e^{i\omega t'}}{i\omega} + \frac{e^{-\gamma t} + e^{-\gamma t'}}{\gamma} \right|. \tag{6.50}$$

The second term within the squared modulus may be suppressed by looking into the limit in which t and t' go to infinity, but their difference $\tau = t - t'$ is kept fixed. It deserves to be noted that such a term leads to an integral that diverges in the limit of an infinite cut-off, as discussed by Breuer and Petruccione (2007). From the physical point of view it means that the central particle can absorb an arbitrary amount of energy and can travel an arbitrary distance within a finite time interval. This singular behavior, known as initial jolts, is clearly a result of the artificial assumption of an uncorrelated initial state. We treat it considering the long-time limit as stated above, where the initial conditions do not affect the motion anymore. It results in

$$\text{MSD}(\tau) = \frac{\gamma\hbar}{m\pi} \int_0^\infty d\omega \frac{\omega}{\omega^2 + \gamma^2} \coth\left(\frac{\hbar\omega}{2k_B T}\right) \frac{4\sin^2(\omega\tau/2)}{\omega^2}. \tag{6.51}$$

For any finite temperature, considering large values of τ one may use the relation

$$\lim_{\tau \to \infty} \frac{\sin(\omega\tau)}{\pi\omega} = \delta(\omega), \tag{6.52}$$

we finally get

$$\text{MSD}(\tau) = 2\frac{k_B T}{m\gamma}t. \tag{6.53}$$

We recover so also in this case the result obtained in Chap. 2 for classical Brownian motion. The approach presented in this chapter permits to highlight the relation existing between the diffusive behavior in Eq. (6.48) and the form of the spectral density: normal diffusion arises for an ohmic spectral density. In the case of the an impurity in a homogeneous gas in one dimension the spectral density shows the cubic behavior in Eq. (6.22), that leads to a position variance which is no longer linear in time. Precisely, Lampo et al. (2017) showed that it is proportional to the square of time (super-diffusion). We can interpret so such a super-diffusive behavior as a witness of memory effects associated to the non-ohmic form.

6.3   Heisenberg Equations for Non-linear Coupling

For sake of completeness we pay a little bit of attention to the form of the equations arising if we consider the non-linear coupling. This problem has already been treated by Barik and Ray (2005). Starting by the Hamiltonian in Eq. (3.1) with the non-linear interaction (4.1), and following the same procedure in Sect. 6.1 one obtains:

$$\dot{x}(t) = p(t), \tag{6.54}$$

and

$$\dot{p}(t) = -U'(x) - f'(x(t)) \int_0^t f'(x(t)) \Gamma(t - t') p(t') dt'$$
$$+ f'(x(t)) \eta(t). \tag{6.55}$$

where $f'(x) = 2x/a$. The quantity

$$\eta(t) = \sum_n \left(\frac{\omega_n}{\left(g_n^{(0)}\right)^2} x_n(0) - f(x(0)) \right) \frac{\left(g_n^{(0)}\right)^2}{\omega_n^2} \cos(\omega_n t)$$
$$+ \sum_n \frac{g_\nu}{\omega_n} p_n(0) \sin(\omega_n t), \tag{6.56}$$

is the noise kernel, where $\omega_n \equiv \omega \epsilon_n$, while

$$\Gamma(t) = \sum_n \left(\frac{g_n}{\omega_n} \right)^2 \cos(\omega_n t), \tag{6.57}$$

is the damping kernel.

In conclusion, the dynamics of the impurity is ruled by the Eqs. (6.54) and (6.55). Their main feature is the position dependence of the damping and the multiplicative noise. This is a consequence of the non-linear coupling in the Hamiltonian, induced by the inhomogeneous character of the gas.

Chapter 7
Conclusions and Perspectives

In this text we presented a short overview of the quantum Brownian motion (QBM) model, and its last extensions. The discussion was driven by the possibility to apply such a model to the Bose polaron, namely the system compound by an impurity in a Bose-Einstein condensate (BEC). This is justified by the fact that the Hamiltonian describing the real aforementioned system may be reduced to that of the QBM model showed in Eq. (3.1). The derivation of the QBM Hamiltonian starting by the impurity-gas one has been developed in Lampo et al. (2017) for a homogeneous BEC, and in Lampo et al. (2018) for an inhomogeneous medium.

In the QBM framework the impurity plays the role of the Brownian particle, while the environment is represented by the Bogoliubov cloud. In this context we treat the impurity as an open quantum system, i.e. we focus on its behavior dealing with that of the Bogoliubov bath as a source of noise, damping and decoherence. Such an approach to the polaron problem is particularly suitable to study the motion of the impurity and in general its dynamics. Precisely we focused on the calculation of the position variance, which has been measured in the experiment of Catani et al. (2012). For this purpose we employed three different techniques:

- Born-Markov master equation (Chap. 3);
- Lindblad equation (Chap. 5);
- Heisenberg equations (Chap. 6).

The Born-Markov equation is a master equation (an equation ruling the temporal evolution of the impurity state, here represented by its reduced density matrix) derived under the Born and Markov hypothesis. The former hypothesis requires that the global impurity-bath state remains separable while the latter assumes that no memory effects occurs. These approximations are usually very well fulfilled at high-temperature and weak bath-impurity coupling, while could fail for example in the zero-temperature limit or when the coupling grows. In fact, the calculation of the position variance leads to violation of the Heisenberg principle (see Fig. 4.2) due to a breaking of the positivity of the Born-Markov equation solution.

© The Author(s), under exclusive license to Springer Nature Switzerland AG 2019
A. Lampo et al., *Quantum Brownian Motion Revisited*, SpringerBriefs
in Physics, https://doi.org/10.1007/978-3-030-16804-9_7

The limitations of the Born-Markov treatment induced us to consider further techniques. We recalled so a Lindblad equation, namely a master equation which structure has been constructed in order to preserve the positivity of its solution, avoiding so the emergence of Heisenberg violations. The Hamiltonian of QBM does not lead to a master equation in a Lindblad form, we propose so a Lindblad equation which form is as close as possible to the original QBM one. Precisely, the Lindblad equation we obtain differs just by an extra-term from the Born-Markov equation derived in Chap. 3. We find that such an extra term (responsible for the Lindblad character of the equation) yields a rotation of the stationary solution in the phase space (see Fig. 5.1).

The fact that the Lindblad equation cannot be derived microscopically by the Hamiltonian of the QBM makes it not so suitable for the study of the polaron. We perform so a description of the system in terms of Heisenberg equations. Such an approach leads to an equation for the impurity which is formally the same to that derived by Langevin for classical Brownian motion. This quantum Langevin-like equation is an important tool to describe the dynamics of the impurity, free of any violation of the Heisenberg principle (see Fig. 6.3). An important feature of this equations is its dependence of the past-history, i.e. it carries out memory effects, vanishing if and only if the bath is ohmic. In the present book we focused on this last case, while the physical consequences of memory effects (super-diffusive form of the position variance) have been discussed in detail in the works of Lampo et al. (2017, 2018).

For each of the aforementioned techniques we payed attention to the possibility to extend them to the situation in which the interaction between bath and impurity presents a non-linear dependence on the position of the latter [see Eq. (4.1)]. This is motivated by the fact that such a non-linear interaction is the natural way to treat the polaron problem. Neberthless, as proved by Lampo ct al. (2017, 2018) a linear approximation is reasonable for realistic values of physical quantities, even though it put some constraints on these.

In these works it has been showed, moreover, so that in order to extend the present analysis to the real system of the polaron one has to consider a different form of the spectral density. The results presented here has been obtained for an ohmic spectral density, while for impurity in a BEC such quantity is super-ohmic. Precisely it is proportional to the third (fourth) power of the frequency for a homogeneous (inhomogeneous) BEC. This implies for instance that the Langevin-like equation for the polaron is affected by memory effects and it yields a position variance proportional to the square of time (super-diffusion). Note that that the fact that the position variance has been measured by Catani et al. (2012) opens the possibility to detect a witness of memory effect on a measurable quantity.

Another important result that can be translated to the polaron is represented by genuine position squeezing, i.e. $\delta_x < 1$, corresponding by high spatial localization of the trapped impurity. Squeezing effect has important applications in quantum metrology and its emergence in a realistic system suggests interesting perspectives for the development of quantum technologies. For instance, it is possible to study new protocols for quantum probing, where one aims to extract properties of a quantum system by means of a probe. Here it is possible to employ the impurity as a probe in order to infer characteristics of the surrounding gas. In particular, motivated by the recent advances in quantum thermometry one could look into the temperature of the gas. So, a possible important outlook of the present book is the possibility of creating a minimally-disturbing method to evaluate temperature fluctuations. This issue has been treated in the work of Mehboudi (2018).

We highlight that the procedure we developed in this work can be exported to other ultracold gases systems. For instance, Efimkin et al. (2016) proved that the dynamics of a bright soliton in a superfluid in one dimension is described by an equation showing the same form of Eq. (6.18). The spectral density for the system is in some circumstances proportional to the third power of the frequency of the Bogoliubov modes, similarly to an impurity in a homogeneous gas. Accordingly, one may apply the techniques presented in Chap. 6 and study the stationary state of the bright soliton. In particular one could wonder if also the bright soliton experiences genuine position squeezing at low-temperature.

Beyond the squeezing effect, the derivation of a motion equation associated to the dynamics of an impurity may be extended to the so-called bipolaron problem, namely a system compound by two particles in a BEC. Particularly, one could look into the very interesting problem of whether the presence of the BEC induces an interaction between the two impurities. This self-induced interaction results to be a very interesting effects in view of the possibility of generating entanglement. Such a problem has been developed in the work of Charalambous et al. (2019).

Another important result of the this manuscript is the detection of the non-Markovian character of the Bose polaron physics. In our framework, non-Markovianity arises because of the super-ohmic form of the spectral density, leading to a non-local in time motion equation for the impurity. Recently, the control and manipulation of non-Markovian effects attracted a lot of attention in the open quantum systems community, especially because of the analysis of the relation with the thermodynamic properties, in the context of the design of new devices in quantum technologies. We already stated that, in some circumstances, non-Markovian dynamics is associated to a back-flow of energy, from the environment to the central system. Moreover, there is a lot of interest in the study of the possibility to extract work form the correlations with the past history. The detection of non-Markovianity in a real physical system, such as an impurity in a BEC, opens the possibility to propose experiments to treat the above mentioned issues.

In general, the possibility to employ quantum Brownian motion to investigate the physics of an impurity in a BEC opens a wide range of chances for testing in a physical system the multitude of effects detected for the present model at an abstract level. For instance, Maniscalco et al. (2006) pointed out the emergence of Zeno and anti-Zeno effect for the quantum Brownian motion. We wonder if is it possible to reveal such an effect for an impurity in a (homogeneous and inhomogeneous) gas. Similarly, one could also propose an experiment with ultracold gases to test the emergence of spectrum broadcast structure for the Bose polaron detected by Tuziemski and Korbicz (2015) and Galve et al. (2016).

Another interesting question concerns the Smoluchowski-Kramers (SK) limit (Smoluchowski 1916; Kramers 1940), which can be considered as a regime of over-damped quantum Brownian motion, or the case where the mass m of the Brownian particle tends to zero. This limit is already highly non-trivial at the classical level, in the presence of the inhomogeneous damping and diffusion, and it requires a careful application of homogenization theory (cf. Hottovy et al. 2012a, 2014; Papanicolaou 1977; Pavliotis and Stuart 2008; Lim et al. 2018). Of course, the theoretical approach here is based on the separation of time scales, and has been in other contexts studied in the theory of classical and quantum stochastic processes (Gardiner 2009; Risken 2012). In particular, the theory of adiabatic elimination has been developed to include the short time non-Markovian "initial slip" effects and the effective long time dynamics of the systems and the bath ("adiabatic drag") (cf. Haake 1982; Haake and Maciej 1983; Haake and Reinhard 1985 and references therein).

The Smoluchowski-Kramers limit was also intensively studied in the contexts of Caldeira-Leggett model and quantum Brownian motion (cf. Maier and Joachim 2010; Ankerhold et al. 2001 and references therein). The problem with this limit is that it corresponds to strong damping, and evidently cannot be described using weak coupling approach that is normally used to derive the master equation from the microscopic model in the Born-Markov approximation. We envisage here two possible and legitimate lines of investigation.

One can forget about the microscopic derivation, and takes the Born-Markov master equation as a starting point. The SK limit corresponds then to setting the spring constant $m\Omega^2$ and friction η to constants, and letting the mass $m \to 0$, so that $\gamma \to \infty$ as $1/m$ and $\Omega \to \infty$ as $1/\sqrt{m}$. The aim is to eliminate the fast variable (the momentum) and to obtain the resulting equation for the position of the Brownian particle; again, the Wigner function formalism is particularly suited for such a task.

More ambitious and physically more sound is the approach in which the microscopic model is treated seriously, and appropriate scalings are introduced at the microscopic level. One can then start, for instance, from the formally exact path integral expression for the reduced dynamics, as pursued by Ankerhold and collaborators Maier and Joachim (2010), Ankerhold et al. (2001), Ankerhold and Hermann (2008). The other possibility is to use a restricted version of the weak coupling assumption, only demanding that the system does not influence the bath, and use the Redfield equation combined with Laplace transform techniques and Zwanzig's approach Zwanzig (1961). To our knowledge, neither of the two above proposed research tasks has been so far realized for the case of inhomogeneous damping.

We remark once more that, for both a homogeneous and inhomogeneous gas, the polaron Hamiltonian shows an interaction term that depends in a non-linear manner on the impurity position. The results we presented have been obtained by performing a linear approximation of the interaction term, which results to be reasonable for realistic values of the physical quantities. Nevertheless the extension of the mechanism we developed to the general case where the impurity-bath coupling retains its original non-linear form remains an important perspective to extend our analysis beyond the constraint associated to the linearization. Of course, one could pursue this path by means of the techniques discussed in the manuscript (master equations, Heisenberg equations), but it also deserve to address that developed in the work of Lim et al. (2018), which allowed to highlight the emergence of a quantum drift due to non-linearity.

Appendix A
Heisenberg Principle for Density Operators

The purpose of this Appendix is to present a self-contained derivation of the Heisenberg uncertainty principle for density operators. We start from the pure state case. Consider an arbitrary state $|\psi\rangle$ and observables A and B. Denoting by $\langle A \rangle$ the mean of the observable A in the state $|\psi\rangle$,

$$\langle A \rangle = \langle \psi | A | \psi \rangle, \tag{A.1}$$

for the variance of A we have

$$\sigma_A^2 = \langle \psi \left| (A - \langle A \rangle)^2 \right| \psi \rangle, \tag{A.2}$$

and similarly for B. For future reference, let us also note that for any real number a,

$$\langle \psi \left| (A - a)^2 \right| \psi \rangle \geq \sigma_A^2. \tag{A.3}$$

The claim we want to prove is

$$\sigma_A^2 \sigma_B^2 \geq \frac{1}{4} \left\langle \frac{[A, B]}{i} \right\rangle^2, \tag{A.4}$$

where the right-hand side contains the mean value of the observable $[A, B]/i$ in the state $|\psi\rangle$. Introducing the vectors

$$|f\rangle = |(A - \langle A \rangle)\psi\rangle \quad \text{and} \quad |g\rangle = |(B - \langle B \rangle)\psi\rangle, \tag{A.5}$$

we have

$$\sigma_A^2 \sigma_B^2 = \langle f | f \rangle \langle g | g \rangle \geq |\langle f | g \rangle|^2, \tag{A.6}$$

where we applied the Cauchy-Schwarz inequality. The right-hand side of the last inequality can be rewritten as

A. Lampo et al., *Quantum Brownian Motion Revisited*, SpringerBriefs in Physics, https://doi.org/10.1007/978-3-030-16804-9

$$|\langle f|g \rangle|^2 = \left(\frac{\langle f|g \rangle + \langle g|f \rangle}{2} \right)^2 + \left(\frac{\langle f|g \rangle - \langle g|f \rangle}{2i} \right)^2, \qquad (A.7)$$

with both terms on the right-hand side non-negative. Rewriting the second term as the square of the mean of the observable $[A, B]/2i$, and leaving the first term out (keeping it would lead to a stronger inequality, called Robertson-Schrödinger inequality), we obtain the desired bound

$$\sigma_A^2 \sigma_B^2 \geq \frac{1}{4} \left\langle \frac{[A, B]}{i} \right\rangle^2, \qquad (A.8)$$

in the pure state case. Now, if $\rho = \sum_j p_j |\phi_j \rangle \langle \phi_j|$ is an arbitrary density operator, with p_j non-negative coefficients summing up to 1, the mean of A in the state ρ equals

$$\langle A \rangle^{(\rho)} = \mathrm{Tr} \, (\rho A). \qquad (A.9)$$

For the variance of A in the state ρ we have

$$\left(\sigma_A^{(\rho)} \right)^2 = \mathrm{Tr} \left[\rho \left(A - \langle A \rangle^{(\rho)} \right)^2 \right], \qquad (A.10)$$

and similarly for B. We thus have

$$\left(\sigma_A^{(\rho)} \right)^2 = \sum_j p_j \left\langle \phi_j \left| \left(A - \langle A \rangle^{(\rho)} \right)^2 \right| \phi_j \right\rangle$$

$$\geq \sum_j p_j \left(\sigma_A^{(\phi_j)} \right)^2. \qquad (A.11)$$

where $\left(\sigma_A^{(\phi_j)} \right)^2$ denotes the variance of A in the state $|\phi_j \rangle$, and in the last step we used inequality Eq. (A.3). Similarly,

$$\left(\sigma_B^{(\rho)} \right)^2 \geq \sum_j p_j \left(\sigma_B^{(\phi_j)} \right)^2, \qquad (A.12)$$

By the (discrete version of) the Cauchy-Schwarz inequality (it is crucial that $p_j \geq 0$ here!) we obtain

$$\left(\sigma_A^{(\rho)} \right)^2 \left(\sigma_B^{(\rho)} \right)^2 \geq \left(\sum_j p_j \sigma_A^{(\phi_j)} \sigma_B^{(\phi_j)} \right)^2, \qquad (A.13)$$

which, using the pure-state version of the uncertainty principle, is bounded from below by

$$\frac{1}{4}\left(\sum_j p_j \left\langle \phi_j \left| \frac{[A, B]}{i} \right| \phi_j \right\rangle\right)^2 = \frac{1}{4}\left(\left\langle \frac{[A, B]}{i} \right\rangle^{(\rho)}\right)^2. \qquad (A.14)$$

This is the desired mixed-state version of the inequality. In the last application of the Cauchy-Schwarz inequality, it is crucial that we are dealing with a density operator, so that the eigenvalues p_j are non-negative.

In the most important case for us, when $A = X$ is the position operator and $B = P$ is the momentum operator, the commutator of A and B is a multiple of identity, $[X, P] = i\hbar I$. The mean value of $\frac{[X,P]}{i}$ in any state is thus equal to $\hbar$ and in particular, for the density operators ρ_t, solving a Lindblad equation we obtain at all times the standard form of the Heisenberg uncertainty principle,

$$\sigma_X^2 \sigma_P^2 \geq \frac{\hbar^2}{4}. \qquad (A.15)$$

Appendix B
Gaussian Approximation

The purpose of this Appendix is to prove that the Gaussian approximation performed on the Lindblad equation for Quadratic QBM preserves its Lindblad form. The demonstration we are about to present considers a Gaussian approximation carried out directly on the master equation, while in Sect. 5.2.2 it has been done on the equations for the moments. As we will show, the two procedures are completely equivalent.

Theorem 1 *For a quadratic Lindblad operator:*

$$L = \tilde{\alpha}a^2 + \tilde{\beta}(a^\dagger)^2 + \tilde{\gamma}a^\dagger a + \tilde{\delta}a + \tilde{\epsilon}a^\dagger + \tilde{\eta} \tag{B.1}$$

the self-consistent Gaussian approximation preserves the Lindblad form (and thus the positivity of ρ and Heisenberg principle).

The annihilation and creation operators are represented respectively by a and $a^\dagger$, while $\tilde{\alpha}, \tilde{\beta}, \tilde{\gamma}, \tilde{\delta}, \tilde{\epsilon}, \tilde{\eta}$ are complex parameters. It is immediate to prove that the Lindblad operator introduced in Eq. (5.33) can be expressed in the form showed in Eq. (B.1). Note that it is possible to assume $\langle a \rangle = 0$, since it just shifts the parameters.

Lemma 1 *The parameter $\tilde{\eta}$ in Eq. (B.1) can be shifted arbitrarily.*

Proof The core of the proof lies in the fact that any additive constant in the definition of the Lindblad operator can be compensated by a re-definition of the Hamiltonian, namely:

$$\frac{\partial \rho}{\partial t} = -\frac{i}{\hbar}[H, \rho] + \mathcal{D}_{L+\Delta\tilde{\eta}}(\rho) \tag{B.2}$$

$$= -\frac{i}{\hbar}[H + \Delta H_{\Delta\tilde{\eta}}, \rho] + \mathcal{D}_L(\rho),$$

where:

A. Lampo et al., *Quantum Brownian Motion Revisited*, SpringerBriefs
in Physics, https://doi.org/10.1007/978-3-030-16804-9

$$\mathcal{D}_L(\rho) = L\rho L^\dagger - L^\dagger L\rho/2 - \rho L^\dagger L/2, \tag{B.3}$$

is the Lindblad dissipator, and:

$$\Delta H_{\Delta\tilde{\eta}} = -\frac{i}{2}[(\Delta\tilde{\eta})L^\dagger - (\Delta\tilde{\eta})^*L], \tag{B.4}$$

with $\Delta\tilde{\eta} \in \mathbb{C}$.

Of course changing of Hamiltonian is allowed, since it just modifies the time dependence of a and $a^\dagger$ in the interaction picture.

Lemma 2 *It is possible to perform the factorization:*

$$L = d_1 d_2, \tag{B.5}$$

with:

$$d_1 = \tilde{A}a + \tilde{B}a^\dagger + \tilde{C}, \tag{B.6}$$
$$d_2 = a + \tilde{D}a^\dagger + \tilde{E}.$$

Proof Comparing Eqs. (B.5) and (B.1), one obtains:

$$\tilde{A} = \tilde{\alpha}, \quad \tilde{A}\tilde{D} + \tilde{B} = \tilde{\gamma}, \quad \tilde{A}\tilde{D} + \tilde{C}\tilde{E} = \tilde{\eta} \tag{B.7}$$
$$\tilde{B}\tilde{D} = \tilde{\beta}, \quad \tilde{A}\tilde{E} + \tilde{C} = \tilde{\delta}, \quad \tilde{B}\tilde{E} + \tilde{C}\tilde{D} = \tilde{\delta},$$

so that

$$\tilde{D} = \tilde{\beta}/\tilde{B}, \quad \tilde{\alpha}\tilde{\beta}/\tilde{B} + \tilde{B} = \tilde{\gamma}. \tag{B.8}$$

provide in general two solutions $\tilde{B}_1$ and $\tilde{B}_2$ for $\tilde{B}$, and

$$\tilde{\alpha}\tilde{E} + \tilde{C} = \tilde{\delta}, \quad \tilde{B}\tilde{E} + (\tilde{\beta}/\tilde{B})\tilde{C} = \tilde{\eta}. \tag{B.9}$$

If we can solve these linear equations for $\tilde{E}$ and $\tilde{C}$, we may plug the solution into $\tilde{\alpha}\tilde{D} + \tilde{C}\tilde{E} = \tilde{\eta}$, and adjust η adequately (which we can do according to Lemma 1).

It is easy to check that the two equations for $\tilde{E}$ and $\tilde{C}$ cannot be solved if $\tilde{B}_1 = \tilde{B}_2 = 0$, which implies $\tilde{\gamma} = 0$ and $\tilde{\alpha}\tilde{\beta} = 0$, i.e. the non-generic case $L = \tilde{\alpha}a^2 + \tilde{\delta}a + \tilde{\epsilon}a^\dagger + \tilde{\eta}$, and the related one with $\tilde{\alpha} = 0$, $\tilde{\beta} \neq 0$. The case $\tilde{\alpha} = \tilde{\beta} = 0$ is trivial, as it corresponds to linear Lindblad operator: for such a case, the Gaussian approximation is not needed, since there exists an exact solution of Gaussian form.

Now we prove the Theorem in the generic case:

Proof of the Theorem We look to the Lindblad dissipator related to the factorized Lindblad operator in Eq. (B.5):

$$\mathcal{D}_L(\rho) = d_1 d_2 \rho d_2^\dagger d_1^\dagger - \frac{1}{2}\{d_2^\dagger d_1^\dagger d_1 d_2, \rho\}. \tag{B.10}$$

In the Gaussian approximation, one replaces pairs of operators by their mean values. "Anomalous" terms generate contributions that may be reabsorbed in the Hamiltonian, such as

$$\langle d_1 d_2 \rangle \left[\rho d_2^\dagger d_1^\dagger - \frac{1}{2}\{d_2^\dagger d_1^\dagger, \rho\} \right] = -\frac{1}{2}\langle d_1 d_2 \rangle [d_2^\dagger d_1^\dagger, \rho], \tag{B.11}$$

and:

$$\langle d_2^\dagger d_1^\dagger \rangle \left[d_1 d_2 \rho - \frac{1}{2}\{d_1 d_2, \rho\} \right] = \frac{1}{2}\langle d_2^\dagger d_1^\dagger \rangle [d_1 d_2, \rho]. \tag{B.12}$$

The non-trivial terms are:

$$\begin{aligned}
&\langle d_2^\dagger d_1 \rangle d_2 \rho d_1^\dagger + \langle d_1^\dagger d_1 \rangle d_2 \rho d_2^\dagger \\
&+\langle d_2^\dagger d_2 \rangle d_1 \rho d_1^\dagger + \langle d_1^\dagger d_2 \rangle d_1 \rho d_2^\dagger \\
&-\left\{ \left(\langle d_2^\dagger d_1 \rangle d_1^\dagger d_2 + \langle d_2^\dagger d_2 \rangle d_1^\dagger d_1 \right), \frac{\rho}{2} \right\} \\
&-\left\{ \left(\langle d_1^\dagger d_1 \rangle d_2^\dagger d_2 + \langle d_1^\dagger d_2 \rangle d_2^\dagger d_1 \right), \frac{\rho}{2} \right\}
\end{aligned} \tag{B.13}$$

The resulting ME has a dissipator of the form:

$$\mathcal{D}_L(\rho) = \sum_{i,j=1,2} \tilde{\Gamma}_{ij} \left(d_i \rho d_j^\dagger - \frac{1}{2}\{d_j^\dagger d_i, \rho\} \right), \tag{B.14}$$

where $\tilde{\Gamma}_{ij} = \langle d_{j'}^\dagger d_{i'} \rangle$, where $1' = 2$ and $2' = 1$. This matrix is evidently positive definite, as follows from the Schwartz inequality, so that the dissipator is again of Lindblad form.

Note that the generalization to many oscillators, many Lindblad operators is straightforward. Note also that the non-generic case is simple to treat. It requires, however, a direct calculation. The quartic Lindblad term in this case is treated as above, while the quadratic one does not need to be touched, since it already describes a Gaussian quantum process. The third order term on the other hand partially vanishes and partially gives contributions to the Hamiltonian in the Gaussian approximation.

The remaining question is whether the approximation that we perform on the level of the ME is the same as the Gaussian de-correlation we performed according to the Wick's theorem prescription at the level of the equations for the moments in Sect. 5.2.2. To illustrate this, we consider an arbitrary operator 0 and we derive the dynamical equations for its average value starting by the ME induced by the superoperator in Eq. (B.10).

The dynamical equation for the average value of an operator O presents the following form:

$$\frac{\partial \langle O \rangle}{\partial t} = h_O^u + h_O^{(1)} - \frac{1}{2}\left(h_O^{(2)} + h_O^{(3)}\right),$$ (B.15)

in which

$$h_O^u = -\frac{i}{\hbar}\mathrm{Tr}\left(O\,[H, \rho]\right)$$ (B.16)

$$h_O^{(1)} = \mathrm{Tr}(Od_1 d_2 \rho d_2^\dagger d_1^\dagger) = \langle d_2^\dagger d_1^\dagger O d_1 d_2 \rangle$$

$$h_O^{(2)} = \mathrm{Tr}(Od_2^\dagger d_1^\dagger d_1 d_2 \rho) = \langle O d_2^\dagger d_1^\dagger d_1 d_2 \rangle$$

$$h_O^{(3)} = \mathrm{Tr}(O\rho d_2^\dagger d_1^\dagger d_1 d_2) = \langle d_2^\dagger d_1^\dagger d_1 d_2 O \rangle.$$

Performing the Gaussian approximation at the level of the equation for the moments means to carry out such an approximation on the average values in Eqs. (B.16),

$$h_O^{(1)} = \mathrm{Tr}(Od_1 d_2 \rho d_2^\dagger d_1^\dagger) = \langle d_2^\dagger d_1^\dagger O d_1 d_2 \rangle,$$ (B.17)

$$\simeq \langle d_2^\dagger d_1^\dagger \rangle \langle O d_1 d_2 \rangle + \langle d_2^\dagger d_1^\dagger O \rangle, \langle d_1 d_2 \rangle - \langle d_2^\dagger d_1^\dagger \rangle \langle O \rangle \langle d_1 d_2 \rangle,$$

$$+ \langle d_2^\dagger d_1 \rangle \langle d_1^\dagger O d_2 \rangle + \langle d_2^\dagger O d_1 \rangle \langle d_1^\dagger d_2 \rangle - \langle d_2^\dagger d_1 \rangle \langle O \rangle \langle d_1^\dagger d_2 \rangle,$$

$$+ \langle d_2^\dagger d_2 \rangle \langle d_1^\dagger O d_1 \rangle + \langle d_2^\dagger O d_2 \rangle \langle d_1^\dagger d_1 \rangle - \langle d_2^\dagger d_2 \rangle \langle O \rangle \langle d_1^\dagger d_1 \rangle,$$

$$h_O^{(2)} = \mathrm{Tr}(Od_2^\dagger d_1^\dagger d_1 d_2 \rho) = \langle O d_2^\dagger d_1^\dagger d_1 d_2 \rangle$$ (B.18)

$$\simeq \langle d_2^\dagger d_1^\dagger \rangle \langle O d_1 d_2 \rangle + \langle O d_2^\dagger d_1^\dagger \rangle \langle d_1 d_2 \rangle - \langle d_2^\dagger d_1^\dagger \rangle \langle O \rangle \langle d_1 d_2 \rangle$$

$$+ \langle d_2^\dagger d_1 \rangle \langle O d_1^\dagger d_2 \rangle + \langle O d_2^\dagger d_1 \rangle \langle d_1^\dagger d_2 \rangle - \langle d_2^\dagger d_1 \rangle \langle O \rangle \langle d_1^\dagger d_2 \rangle$$

$$+ \langle d_2^\dagger d_2 \rangle \langle O d_1^\dagger d_1 \rangle + \langle O d_2^\dagger d_2 \rangle \langle d_1^\dagger d_1 \rangle - \langle d_2^\dagger d_2 \rangle \langle O \rangle \langle d_1^\dagger d_1 \rangle,$$

$$h_O^{(3)} = \mathrm{Tr}(O\rho d_2^\dagger d_1^\dagger d_1 d_2) = \langle d_2^\dagger d_1^\dagger d_1 d_2 O \rangle$$ (B.19)

$$\simeq \langle d_2^\dagger d_1^\dagger \rangle \langle d_1 d_2 O \rangle + \langle d_2^\dagger d_1^\dagger O \rangle \langle d_1 d_2 \rangle - \langle d_2^\dagger d_1^\dagger \rangle \langle O \rangle \langle d_1 d_2 \rangle$$

$$+ \langle d_2^\dagger d_1 \rangle \langle d_1^\dagger d_2 O \rangle + \langle d_2^\dagger d_1 O \rangle \langle d_1^\dagger d_2 \rangle - \langle d_2^\dagger d_1 \rangle \langle O \rangle \langle d_1^\dagger d_2 \rangle$$

$$+ \langle d_2^\dagger d_2 \rangle \langle d_1^\dagger d_1 O \rangle + \langle d_2^\dagger d_2 O \rangle \langle d_1^\dagger d_1 \rangle - \langle d_2^\dagger d_2 \rangle \langle O O \rangle \langle d_1^\dagger d_1 \rangle.$$

It is now tedious but easy to check that replacing the expressions in Eqs. (B.17–B.19) in Eq. (B.15) we get the dynamical equations generated by the terms in Eqs. (B.11–B.13), obtained by performing the Gaussian approximation on the master equation related to a dissipator in Eq. (B.10). This proves that performing the Gaussian approximation at the level of the master equation is equivalent to doing it at the level of the equations for the moments of an observable. Note that the equations resulting by this approximation will always admit a Gaussian solution, although it is not guaranteed that the latter is stationary.

The demonstration we developed holds for Lindblad operators which are quadratic in the creation and annihilation operators. This case covers the situation studied in Sect. (5.2), but it is not the most general one. In fact, one could consider also Lindblad equations with Lindblad operators containing higher powers of creation and annihilation operators. Extending the proof we presented to this general case is an interesting perspective that we reserve for future works.

References

Abbot, Davies, Pati: Quantum Aspects of the Life. World Scientific, Oxford (2008). http://www.worldscientific.com/worldscibooks/10.1142/p581; ISBN: 978-1-84816-253-2

Alexandrov, A.S., Devreese, J.T.: Advances in Polaron Physics. Springer Series in Solid-State Sciences. Springer (2009). http://books.google.es/books?id=EI0Hql-9oY8C; ISBN: 9783642018961

Ankerhold, J., Grabert, H.: Erratum: strong friction limit in quantum mechanics: the quantum Smoluchowski equation [Phys. Rev. Lett. 87, 086802 (2001)]. Phys. Rev. Lett. **101**(11), 119903 (2018). https://doi.org/10.1103/PhysRevLett.101.119903. http://link.aps.org/doi/10.1103/PhysRevLett.101.119903

Ankerhold, J., Pechukas, P., Grabert, H.: Strong friction limit in quantum mechanics: the quantum Smoluchowski equation. Phys. Rev. Lett. **87**(8), 086802 (2001). https://doi.org/10.1103/PhysRevLett.87.086802. http://link.aps.org/doi/10.1103/PhysRevLett.87.086802

Ardila, L.A., Pena, Giorgini, S.: Impurity in a Bose-Einstein condensate: study of the attractive and repulsive branch using quantum Monte Carlo methods. Phys. Rev. A **92**(3), 033612 (2015). https://doi.org/10.1103/PhysRevA.92.033612

Ardila, L.A., Pena, Giorgini, S.: Bose polaron problem: effect of mass imbalance on binding energy. Phys. Rev. A **94**(6), 063640 (2016). https://doi.org/10.1103/PhysRevA.94.063640. https://link.aps.org/doi/10.1103/PhysRevA.94.063640

Bakker, G.J., et al.: Lateral mobility of individual integrin nanoclusters orchestrates the onset for leukocyte adhesion. Proc. Nat. Acad. Sci. **109**(13), 4869–4874 (2012). https://doi.org/10.1073/pnas.1116425109. http://www.pnas.org/content/109/13/4869.abstract

Banerjee, S., Ghosh, R.: General quantum Brownian motion with initially correlated and nonlinearly coupled environment. Phys. Rev. E **67**(5), 056120 (2003). https://doi.org/10.1103/PhysRevE.67.056120. http://link.aps.org/doi/10.1103/PhysRevE.67.056120

Barik, D., Shankar Ray, D.: Quantum state-dependent diffusion and multiplicative noise: a microscopic approach. J. Stat. Phys. **120**(1), 339–365 (2005). ISSN: 1572-9613. https://doi.org/10.1007/s10955-005-5251-y. https://doi.org/10.1007/s10955-005-5251-y

Benjamin, D., Demler, E.: Variational polaron method for Bose-Bose mixtures. Phys. Rev. A **89**(3), 033615 (2014). https://doi.org/10.1103/PhysRevA.89.033615. https://link.aps.org/doi/10.1103/PhysRevA.89.033615

Bloch, I., Dalibard, J., Zwerger, W.: Manybody physics with ultracold gases. Rev. Mod. Phys. **80**(3), 885–964 (2008). https://doi.org/10.1103/RevModPhys.80.885. http://link.aps.org/doi/10.1103/RevModPhys.80.885

Blume-Kohout, R., Zurek, W.H.: Quantum Darwinism in quantum Brownian motion. Phys. Rev. Lett. **101**(24), 240405 (2008). https://doi.org/10.1103/PhysRevLett.101.240405. http://link.aps.org/doi/10.1103/PhysRevLett.101.240405

Bouchaud, J.-P., Georges, A.: Anomalous diffusion in disordered media: Statistical mechanisms, models and physical applications. Physics Reports **195**, 127–293 (1990). https://doi.org/10.1016/0370-1573(90)90099-N

Boyanovsky, D., Jasnow, D.: Heisenberg-Langevin versus quantum master equation. Phys. Rev. A **96**(6), 062108 (2017). https://doi.org/10.1103/PhysRevA.96.062108. https://link.aps.org/doi/10.1103/PhysRevA.96.062108

Brettschneider, T., et al.: Force measurement in the presence of Brownian noise: equilibrium-distribution method versus drift method. Phys. Rev. E **83**, 041113 (2011)

Breuer, H.P., Petruccione, F.: The Theory of Open Quantum Systems. OUP, Oxford (2007). http://books.google.es/books?id=DkcJPwAACAAJ. ISBN 9780199213900

Brown, R.: A brief account of microscopical observations made in the months of June, July and August 1827, on the particles contained in the pollen of plants; and on the general existence of active molecules in organic and inorganic bodies. Philos. Mag. **4**(21), 161–173 (1828). https://doi.org/10.1080/14786442808674769. https://doi.org/10.1080/14786442808674769

Brown, R.: Additional remarks on active molecules. Philos. Mag. **6**(33), 161–166 (1829). https://doi.org/10.1080/14786442908675115. https://doi.org/10.1080/14786442908675115

Brun, T.A.: Quasiclassical equations of motion for nonlinear Brownian systems. Phys. Rev. D **47**(8), 3383–3393 (1993). https://doi.org/10.1103/PhysRevD.47.3383. http://link.aps.org/doi/10.1103/PhysRevD.47.3383

Caldeira, A.O., Leggett, A.J.: Path integral approach to quantum Brownian motion. Phys. A: Stat. Mech. Appl. **121**(3), 587–616 (1983a). ISSN: 0378-4371. https://doi.org/10.1016/0378-4371(83)90013-4. http://www.sciencedirect.com/science/article/pii/0378437183900134

Caldeira, A.O., Leggett, A.J.: Quantum tunnelling in a dissipative system. Ann. Phys. **149**(2), 374–456 (1983b). ISSN: 0003-4916. https://doi.org/10.1016/0003-4916(83)90202-6. http://www.sciencedirect.com/science/article/pii/0003491683902026

Castelnovo, C., Caux, J.-S., Simon, S.H.: Driven impurity in an ultracold one-dimensional Bose gas with intermediate interaction strength. Phys. Rev. A **93**(1), 013613 (2016). https://doi.org/10.1103/PhysRevA.93.013613. https://link.aps.org/doi/10.1103/PhysRevA.93.013613

Catani, J., et al.: Quantum dynamics of impurities in a one-dimensional Bose gas. Phys. Rev. A **85**(2), 023623 (2012). https://doi.org/10.1103/PhysRevA.85.023623. http://link.aps.org/doi/10.1103/PhysRevA.85.023623

Charalambous, C. et al.: Two distinguishable impurities in BEC: squeezing and entanglement of two Bose polarons. SciPost Phys. **6**(1), 10 (2019). https://doi.org/10.21468/SciPostPhys.6.1.010. https://scipost.org/10.21468/SciPostPhys.6.1.010

Cherstvy, A.G., Ralf, M.: Population splitting, trapping, and non-ergodicity in heterogeneous diffusion processes. Phys. Chem. Chem. Phys. **15**(46), 20220–20235 (2013). https://doi.org/10.1039/C3CP53056F. https://doi.org/10.1039/C3CP53056F

Christensen, R.S., Levinsen, J., Bruun, G.M.: Quasiparticle properties of a mobile impurity in a Bose-Einstein condensate. Phys. Rev. Lett. **115**(16), 160401 (2015). https://doi.org/10.1103/PhysRevLett.115.160401. https://link.aps.org/doi/10.1103/PhysRevLett.115.160401

Cisse, I.I., et al.: Real-time dynamics of RNA polymerase II clustering in live human cells. Science **341**(6146), 664–667 (2013). https://doi.org/10.1126/science.1239053. http://www.sciencemag.org/content/341/6146/664.abstract

Côté, R., Kharchenko, V., Lukin, M.D.: Mesoscopic molecular ions in Bose-Einstein condensates. Phys. Rev. Lett. **89**(9), 093001 (2002). https://doi.org/10.1103/PhysRevLett.89.093001. http://link.aps.org/doi/10.1103/PhysRevLett.89.093001

Cucchietti, F.M., Timmermans, E.: Strong-coupling polarons in Dilute Gas Bose-Einstein condensates. Phys. Rev. Lett. **96**(21), 210401 (2006). https://doi.org/10.1103/PhysRevLett.96.210401. http://link.aps.org/doi/10.1103/PhysRevLett.96.210401

Diósi, L.: On high-temperature Markovian equation for quantum Brownian motion. EPL (Europhysics Letters) **22**(1), 1 (1993). http://stacks.iop.org/0295-5075/22/i=1/a=001

Dykman, M.I., Krivoglaz, M.A. Spectral distribution of nonlinear oscillators with nonlinear friction due to a medium. Phys. Status Solidi (B) **68**(1), 111–123 (1975). ISSN: 1521-3951. https://doi.org/10.1002/pssb.2220680109. http://dx.doi.org/10.1002/pssb.2220680109

Efimkin, D.K., Hofmann, J., Galitski, V.: Non-Markovian quantum friction of bright solitons in superfluids. Phys. Rev. Lett. **116**(22), 225301 (2016). https://doi.org/10.1103/PhysRevLett.116.225301. http://link.aps.org/doi/10.1103/PhysRevLett.116.225301

Einstein, A.: Die von der Molekularkinetischen Theorie von Wärme geforderte Bewegung von in ruhenden Flüssigktein suspendierten Teilchen. Ann. Phys. **17**(549) (1905)

Einstein, A.: Eine neue Bestimmung der Molekuldimensionen. Ann. Phys. **19**(289) (1906a)

Einstein, A.: Zur Theorie der Brownschen Bewegung. Ann. Phys. **19**(371) (1906b)

Einstein, A.: Theoretische Bemerkungen über die Brownsche Bewegung. Zeit. f. Electrochem **13**(371) (1907)

Einstein, A.: Elementaire Theorie der Brownsche Bewegung. Zeit. f. Electrochem **14**(235) (1908)

Exner, F.M.: Notiz zu Brown's Molecularbewegung. Annalen der Physik **307**(8), 843–847. ISSN: 1521-3889. https://doi.org/10.1002/andp.19003070813. http://dx.doi.org/10.1002/andp.19003070813

Fleming, C.H., Cummings, N.I.: Accuracy of perturbative master equations. Phys. Rev. E **83**(3), 031117 (2011). https://doi.org/10.1103/PhysRevE.83.031117. http://link.aps.org/doi/10.1103/PhysRevE.83.031117

Ford, G.W., O'Connell, R.F.: Comment on dissipative quantum dynamics with a lindblad functional. Phys. Rev. Lett. **82**(16), 3376–3376 (1999). https://doi.org/10.1103/PhysRevLett.82.3376. http://link.aps.org/doi/10.1103/PhysRevLett.82.3376

Fröhlich, H.: Electrons in lattice fields. Adv. Phys.**3**(11), 325 (1954). http://www.tandfonline.com/doi/abs/10.1080/00018735400101213

Fukuhara, T., et al.: Quantum dynamics of a mobile spin impurity. Nat. Phys. **9**, 235–241 (2013). https://doi.org/10.1038/nphys2561

Galve, F., Zambrini, R., Maniscalco, S.: Non-Markovianity hinders quantum Darwinism. Scientific Reports **6**, (2016). https://doi.org/10.1038/srep19607. http://www.nature.com/articles/srep19607

Gao, S.: Dissipative quantum dynamics with a Lindblad functional. Phys. Rev. Lett. **79**(17), 3101–3104 (1997). https://doi.org/10.1103/PhysRevLett. 79 3101.http://link.aps.org/doi/10.1103/PhysRevLett.79.3101

Gao, S.: Gao replies. Phys. Rev. Lett. **80**(25), 5703–5703 (1998). https://doi.org/10.1103/PhysRevLett.80.5703. http://link.aps.org/doi/10.1103/PhysRevLett.80.5703

Gao, S.: Gao replies. Phys. Rev. Lett. **82**(16), 3377–3377 (1999). https://doi.org/10.1103/PhysRevLett.82.3377. http://link.aps.org/doi/10.1103/PhysRevLett.82.3377

Gardiner, C., Zoller, P.: Quantum Noise: A Handbook of Markovian and Non-Markovian Quantum Stochastic Methods with Applications to Quantum Optics. Springer Series in Synergetics. Springer, Berlin (2004)

Gardiner, C.W.: Stochastic Methods: A Handbook for the Natural and Social Sciences, vol. 13. Springer Series in Synergetics. Springer, Heidelberg (2009)

Gil, G.M., et al.: Transient subdiffusion from an Ising environment. Phys. Rev. E **96**(5), 052140 (2017). https://doi.org/10.1103/PhysRevE.96.052140. https://link.aps.org/doi/10.1103/PhysRevE.96.052140

Golding, I., Cox, E.C.: Physical nature of bacterial cytoplasm. Phys. Rev. Lett. **96**(9), 098102 (2006). https://doi.org/10.1103/PhysRevLett.96.098102. http://link.aps.org/doi/10.1103/PhysRevLett.96.098102

Gorini, V., Kossakowski, A., Sudarshan, E.C.G.: Completely positive dynamical semigroups of N-level systems. J. Math. Phys. **17**, 821–825 (1976). https://doi.org/10.1063/1.522979

Grabert, H., Schramm, P., Ingold, G-L.: Quantum Brownian motion: the functional integral approach. Phys. Rep. **168**(3), 115–207 (1988). ISSN: 0370-1573. https://doi.org/10.1016/0370-1573(88)90023-3. http://www.sciencedirect.com/science/article/pii/0370157388900233

Gröblacher, S et al.: Observation of non-Markovian micromechanical Brownian motion. Nat. Commun. **6**, 7606 (2015). ISSN: 2041-1723. https://doi.org/10.1038/ncomms8606. http://europepmc.org/articles/PMC4525213

Grusdt, F., Demler, E.: New theoretical approaches to Bose polarons. (2016) In: arXiv:1510.04934. https://arxiv.org/pdf/1510.04934.pdf

Grusdt, F. et al.: Bloch oscillations of bosonic lattice polarons. (2014a) In: arXiv: 1410.1513

Grusdt, F. et al.: Renormalization group approach to the Fröhlich polaron model: application to impurity-BEC problem. (2014b) In: arXiv: 1410.2203

Grusdt, F., Fleischhauer, M.: Tunable Polarons of slow- light polaritons in a two-dimensional Bose-Einstein condensate. Phys. Rev. Lett. **116**(5), 053602 (2016). https://doi.org/10.1103/PhysRevLett.116.053602. https://link.aps.org/doi/10.1103/PhysRevLett.116.053602

Haake, F., Lewenstein, M., Reibold, R.: Adiabatic drag and initial slip for random processes with slow and fast variables. In: Accardi, L., vonWaldenfels, W. (eds.) Lecture Note in Mathematics: Quantum Probability and Applications II. Springer, Heidelberg (1985)

Haake, F.: Systematic adiabatic elimination for stochastic processes. English. In: Zeitschrift für Physik B Condensed Matter **48**(1), 31–35 (1982). ISSN: 0722-3277. https://doi.org/10.1007/BF02026425. http://dx.doi.org/10.1007/BF02026425

Haake, F., Lewenstein, M.: Adiabatic drag and initial slip in random processes. Phys. Rev. A **28**(6), 3606–3612 (1983). https://doi.org/10.1103/PhysRevA.28.3606. http://link.aps.org/doi/10.1103/PhysRevA.28.3606

Haake, F., Reibold, R.: Strong damping and low-temperature anomalies for the harmonic oscillator. Phys. Rev. A **29**, 3208 (1984)

Haake, F., Reibold, R.: Strong damping and low-temperature anomalies for the harmonic oscillator. Phys. Rev. A **32**(4), 2462–2475 (1985). https://doi.org/10.1103/PhysRevA.32.2462. http://link.aps.org/doi/10.1103/PhysRevA.32.2462

Haus, J.W., Kehr, K.: Diffusion in regular and disordered lattices. Phys. Rep. **150**, 263–406 (1987)

Havlin, S., Ben-Avraham, D.: Diffusion in disordered media. Adv. Phys. **36**(6), 695–798 (1987). https://doi.org/10.1080/00018738700101072. http://www.tandfonline.com/doi/abs/10.1080/00018738700101072

Höfling, F., Franosch, T.: Anomalous transport in the crowded world of biological cells. Rep. Prog. Phys. **76**(4), 046602 (2013). https://doi.org/10.1088/0034-4885/76/4/046602

Hottovy, S., Volpe, G., Wehr, J.: Noise-induced drift in stochastic differential equations with arbitrary friction and diffusion in the Smoluchowski-Kramers limit. J. Stat. Phys. **146**, 762 (2012a)

Hottovy, S., Volpe, G., Wehr, J.: Thermophoresis of Brownian particles driven by coloured noise. EPL **99**, 60002 (2012b)

Hottovy, S., Volpe, G., Wehr, J.: The Smoluchowski-Kramers limit of stochastic differential equations with arbitrary state-dependent friction. Preprint (Comm. Math. Phys., in press) (2014). arXiv:1404.2330

Hu, B.L., Paz, J.P., Zhang, Y.: Quantum Brownian motion in a general environment: Exact master equation with nonlocal dissipation and colored noise. Phys. Rev. D **45**(8), 2843–2861 (1992). https://doi.org/10.1103/PhysRevD.45.2843. http://link.aps.org/doi/10.1103/PhysRevD.45.2843

Hu, B.L., Paz, J.P., Zhang, Y.: Quantum Brownian motion in a general environment. II. Nonlinear coupling and perturbative approach. Phys. Rev. D **47**(4), 1576–1594 (1993). https://doi.org/10.1103/PhysRevD.47.1576. http://link.aps.org/doi/10.1103/PhysRevD.47.1576

Hu, M.-G., et al.: Bose polarons in the strongly interacting regime. Phys. Rev. Lett. **117**(5), 055301 (2016). https://doi.org/10.1103/PhysRevLett.117.055301. https://link.aps.org/doi/10.1103/PhysRevLett.117.055301

Isar, A., et al.: Open quantum systems. Int. J. Mod. Phys. E **03**(02), 635–714 (1994). https://doi.org/10.1142/S0218301394000164. http://www.worldscientific.com/doi/abs/10.1142/S0218301394000164

Jeon, J.-H., et al.: In vivo anomalous diffusion and weak ergodicity breaking of lipid granules. Phys. Rev. Lett. **106**(4), 048103 (2011). https://doi.org/10.1103/PhysRevLett.106.048103. http://link.aps.org/doi/10.1103/PhysRevLett.106.048103

Jørgensen, N.B., et al.: Observation of attractive and repulsive polarons in a Bose-Einstein condensate. Phys. Rev. Lett. **117**(5), 055302 (2016). https://doi.org/10.1103/PhysRevLett.117.055302. https://link.aps.org/doi/10.1103/PhysRevLett.117.055302

Kirton, P.G., et al.: Quantum current noise from a Born-Markov master equation. Phys. Rev. B **86**(8), 081305 (2012). https://doi.org/10.1103/PhysRevB.86.081305. https://link.aps.org/doi/10.1103/PhysRevB.86.081305

Klafter, J., Sokolov, I.M.: First Steps in Random Walks. Oxford University Press, Oxford (2011)

Kohstall, C. et al. Metastability and coherence of repulsive polarons in a strongly interacting Fermi mixture. Nature **485**, 615–618 (2012). ISSN: 0028-0836; http://dx.doi.org/10.1038/nature11065

Koschorreck, M., et al.: Attractive and repulsive Fermi polarons in two dimensions. Nature **485**, 619–622 (2012). https://doi.org/10.1038/nature11151

Kramers, H.A.: Brownian motion in a field of force and the diffusion model of chemical reactions. Physica **7**(4), 284–304 (1940). ISSN: 0031-8914. https://doi.org/10.1016/S0031-8914(40)90098-2. http://www.sciencedirect.com/science/article/pii/S0031891440900982

Krinner, S. et al.: Direct observation of fragmentation in a disordered, strongly interacting fermi gas (2013). arXiv: 1311.5174

Kumar, J., Sinha, S., Sreeram, P.A.: Dissipative dynamics of a harmonic oscillator: a nonperturbative approach. Phys. Rev. E **80**(3), 031130 (2009). https://doi.org/10.1103/PhysRevE.80.031130. http://link.aps.org/doi/10.1103/PhysRevE.80.031130

Kusumi, A., et al.: Dynamic organizing principles of the plasma membrane that regulate signal transduction: commemorating the fortieth anniversary of Singer and Nicolson's Fluid-Mosaic model. Annu. Rev. Cell Dev. Biol. **28**(1), 215–250 (2012). https://doi.org/10.1146/annurev-cellbio-100809-151736. http://www.annualreviews.org/doi/abs/10.1146/annurev-cellbio-100809-151736

Lampo, A., et al.: Lindblad model of quantum Brownian motion. Phys. Rev. A **94**(4), 042123 (2016). https://doi.org/10.1103/PhysRevA.94.042123. http://link.aps.org/doi/10.1103/PhysRevA.94.042123

Lampo, A. et al.: Bose polaron as an instance of quantum Brownian motion. Quantum **1**, 30 (2017). ISSN: 2521-327X; https://doi.org/10.22331/q-2017-09-27-30. https://doi.org/10.22331/q-2017-09-27-30

Lampo, A., et al.: Non-Markovian polaron dynamics in a trapped Bose-Einstein condensate. Phys. Rev. A **98**(6), 063630 (2018). https://doi.org/10.1103/PhysRevA.98.063630. https://link.aps.org/doi/10.1103/PhysRevA.98.063630

Lan, Z., Lobo, C.: A single impurity in an ideal atomic Fermi gas: current understanding and some open problems. J. Indian I. Sci. **94**, 179 (2014). http://journal.library.iisc.ernet.in/index.php/iisc/search/advancedResults

Landau, L.D., Pekar, S.I.: Effective mass of a polaron. Zh. Eksp. Teor. Fiz. (1948). http://www.ujp.bitp.kiev.ua/files/journals/53/si/53SI15p.pdf

Landauer, R.: Spatial variation of currents and fields due to localized scatterers in metallic conduction. I.B.M. J. Res. Develop. **149**(1), 223 (1957)

Langevin, P.: Sur la theorie de mouvement brownien. C. R. Acad. Sci. Paris. **146**, 530 (1908)

Leggett, A.J., et al.: Dynamics of the dissipative two-state system. Rev. Mod. Phys. **59**(1), 1–85 (1987). https://doi.org/10.1103/RevModPhys.59.1. http://link.aps.org/doi/10.1103/RevModPhys.59.1

Levinsen, J., Parish, M.M.: Strongly interacting two-dimensional Fermi gases (2014). In: arXiv: 1408.2737

Levinsen, J., Parish, M.M., Bruun, G.M.: Impurity in a Bose-Einstein condensate and the efimov effect. Phys. Rev. Lett. **115**(12), 125302 (2015). https://doi.org/10.1103/PhysRevLett.115.125302. https://link.aps.org/doi/10.1103/PhysRevLett.115.125302

Lewenstein, M., Rzazewski, K.: Collective radiation and the near-zone field. J. Phys. A: Math. Gen. **13**(2), 743 (1980). http://stacks.iop.org/0305-4470/13/i=2/a=035

Lewenstein, M., Sanpera, A., Ahufinger, V.: Ultracold Atoms in Optical Lattices: Simulating Quantum Many-body Systems. OUP, Oxford (2012)

Lim, S.H. et al.: On the small mass limit of quantum brownian motion with inhomogeneous damping and diffusion. J. Stat. Phys. **170**(2), 351–377 (2018). ISSN: 1572-9613; https://doi.org/10.1007/s10955-017-1907-7. https://doi.org/10.1007/s10955-017-1907-7

Lindblad, G.: On the generators of quantum dynamical semigroups. Comm. Math. Phys. **48**(2), 119–130 (1976a). http://projecteuclid.org/euclid.cmp/1103899849

Lindblad, G.: Brownian motion of a quantum harmonic oscillator. Rep. Math. Phys. **10**(3), 393–406 In:. ISSN: 0034-4877; https://doi.org/10.1016/0034-4877(76)90029-X. http://www.sciencedirect.com/science/article/pii/003448777690029X

Maier, S.A., Ankerhold, J.: Quantum smoluchowski equation: a systematic study. Phys. Rev. E **81**(2), 021107 (2010). https://doi.org/10.1103/PhysRevE.81.021107. http://link.aps.org/doi/10.1103/PhysRevE.81.021107

Maniscalco, S., et al.: Lindblad and non-Lindblad type dynamics of a quantum Brownian particle. Phys. Rev. A **70**(3), 032113 (2004). https://doi.org/10.1103/PhysRevA.70.032113. http://link.aps.org/doi/10.1103/PhysRevA.70.032113

Maniscalco, S., Piilo, J., Suominen, K.-A.: Zeno and Anti-Zeno effects for quantum Brownian motion. Phys. Rev. Lett. **97**(13), 130402 (2006). https://doi.org/10.1103/PhysRevLett.97.130402. http://link.aps.org/doi/10.1103/PhysRevLett.97.130402

Manzo, C. et al.: Weak ergodicity breaking of receptor motion in living cells stemming from random diffusivity (2014). In: arXiv: 1407.2552

Markov, A.A.: Investigation of a remarkable case of dependent trials. Izv. Akad. Nauk St. Petersburg **6**, 61 (1907)

Marshall, W., et al.: Towards quantum superpositions of a mirror. Phys. Rev. Lett. **91**(13), 130401 (2003). https://doi.org/10.1103/PhysRevLett.91.130401. http://link.aps.org/doi/10.1103/PhysRevLett.91.130401

Massignan, P., Pethick, C.J., Smith, H.: Static properties of positive ions in atomic Bose-Einstein condensates. Phys. Rev. A **71**(2), 023606 (2005). https://doi.org/10.1103/PhysRevA.71.023606. http://link.aps.org/doi/10.1103/PhysRevA.71.023606

Massignan, P., et al.: Nonergodic subdiffusion from Brownian motion in an inhomogeneous medium. Phys. Rev. Lett. **112**(15), 150603 (2014). https://doi.org/10.1103/PhysRevLett.112.150603. http://link.aps.org/doi/10.1103/PhysRevLett.112.150603

Massignan, P., Zaccanti, M., Bruun, G.M.: Polarons, dressed molecules and itinerant ferromagnetism in ultracold Fermi gases. Rep. Prog. Phys. **77**(3), 034401 (2014). http://stacks.iop.org/0034-4885/77/i=3/a=034401

Massignan, P., et al.: Quantum Brownian motion with inhomogeneous damping and diffusion. Phys. Rev. A **91**(3), 033627 (2015). https://doi.org/10.1103/PhysRevA.91.033627. http://link.aps.org/doi/10.1103/PhysRevA.91.033627

Mazo, R.M.: Brownian Motion: Fluctuations, Dynamics, and Applications. Clarendon Press (2002). ISBN 9780198515678; http://eu.wiley.com/WileyCDA/WileyTitle/productCd-0471257095.html

McDaniel, A. et al.: An SDE approximation for stochastic differential delay equations with colored state-dependent noise (2014). In: arXiv: 1406.7287

Mehboudi, M. et al.: Using polarons for sub-nK quantum non-demolition thermometry in a Bose-Einstein condensate. Phys. Rev. Lett. (2018) https://journals.aps.org/prl/accepted/bd079Yd8H831b76754fe8754887ef34ec976c3047

Metzler, R., Klafter, J.: The restaurant at the end of the random walk: recent developments in the description of anomalous transport by fractional dynamics. J. Phys. A: Math. Gen. **37**(31), R161 (2004). http://stacks.iop.org/0305-4470/37/i=31/a=R01

Metzler, R., et al.: Anomalous diffusion models and their properties: non-stationarity, non-ergodicity, and ageing at the centenary of single particle tracking. Phys. Chem. Chem. Phys. **16**(44), 24128–24164 (2014). https://doi.org/10.1039/C4CP03465A. http://dx.doi.org/10.1039/C4CP03465A

Montroll, E.W., Weiss, G.H.: Random walks on lattices. II. J. Math. Phys. **6**(2), pp. 167–181. https://doi.org/10.1063/1.1704269. http://scitation.aip.org/content/aip/journal/jmp/6/2/10.1063/1.1704269

Moy, G.M., Hope, J.J., Savage, C.M.: Born and Markov approximations for atom lasers. Phys. Rev. A **59**(1), 667–675 (1999). https://doi.org/10.1103/PhysRevA.59.667. https://link.aps.org/doi/10.1103/PhysRevA.59.667

Newburgh, R., Peidle, J., Rueckner, W.: Einstein, Perrin, and the reality of atoms: 1905 revisited. Am. J. Phys. **74**, 478 (2006). https://doi.org/10.1119/1.2188962. http://dx.doi.org/10.1119/1.2188962

Palzer, S., et al.: Quantum transport through a Tonks-Girardeau gas. Phys. Rev. Lett. **103**(15), 150601 (2009). https://doi.org/10.1103/PhysRevLett.103.150601. http://link.aps.org/doi/10.1103/PhysRevLett.103.150601

Papanicolaou, G.C.: Modern modeling of continuum phenomena (Ninth Summer Sem. Appl. Math., Rensselaer Polytech. Inst., Troy, N.Y., 1975). Vol. 16. Lect. in Appl. Math. Amer. Math. Soc. 109–147 (1977)

Pavliotis, G.A., Stuart, A.M.: Multiscale Methods, vol. 53. Texts in Applied Mathematics. Springer, New York (2008)

Perrin, J.B.: La loi de Stokes et le mouvment brownien. C. R. Acad. Sci. Paris **147**, 475 (1908a)

Perrin, J.B.: L'origen de mouvement brownien. C. R. Acad. Sci. Paris **147**, 530 (1908b)

Perrin, J.B.: Mouvment brownien et realité moleculair. Ann. de Chim.et de Phys. **18**(1) (1909)

Pesce, G. et al.: Dynamical effects of multiplicative feedback on a noisy system. In: Preprint (2012). arXiv:1206.6271

Rath, S.P., Schmidt, R.: Field-theoretical study of the Bose polaron. Phys. Rev. A **88**(5), 053632 (2013). https://doi.org/10.1103/PhysRevA.88.053632. http://link.aps.org/doi/10.1103/PhysRevA.88.053632

Rentrop, T., et al.: Observation of the Phononic Lamb Shift with a Synthetic Vacuum. Phys. Rev. X **6**(4), 041041 (2016). https://doi.org/10.1103/PhysRevX.6.041041. https://link.aps.org/doi/10.1103/PhysRevX.6.041041

Risken, H.: The Fokker-Planck Equation: Methods of Solution and Applications, vol. 18. Springer Series in Synergetics. Springer, Heidelberg (2012). ISBN: 9783642968099; http://books.google.es/books?id=ThNnMQEACAAJ

Rivas, Á., Huelga, S.F.: Open Quantum Syst. (2012). https://doi.org/10.1007/978-3-642-23354-8

Robinson, N.J., Caux, J.-S., Konik, R.M.: Motion of a distinguishable impurity in the Bose Gas: arrested expansion without a lattice and impurity snaking. Phys. Rev. Lett. **116**(14), 145302 (2016). https://doi.org/10.1103/PhysRevLett.116.145302. https://link.aps.org/doi/10.1103/PhysRevLett.116.145302

Rzazewski, K., Zakowicz, W.: On the interaction of harmonic oscillators with the radiation field. Il Nuovo Cimento B (1971–1996) **1**(2), 111–122 (1971). ISSN: 1826-9877; https://doi.org/10.1007/BF02815271. https://doi.org/10.1007/BF02815271

Rzazewski, K., Zakowicz, W.: Initial value problem and causality of radiating oscillator. J. Phys. A: Math. Gen. **9**(7), 1159 (1976). http://stacks.iop.org/0305-4470/9/i=7/a=018

Rzazewski, K., Zakowicz, W.: Initial value problem for two oscillators interacting with electromagnetic field. J. Math. Phys. **21**(2), 378–388 (1980). https://doi.org/10.1063/1.524426. https://doi.org/10.1063/1.524426

Saxton, M.J.: Lateral diffusion in an archipelago. Single-particle diffusion. Biophys. J. **64**(6), 1766–1780 (1993). ISSN: 0006-3495; https://doi.org/10.1016/S0006-3495(93)81548-0. http://www.pubmedcentral.nih.gov/articlerender.fcgi?artid=1262511&tool=pmcentrez&rendertype=abstract

Scher, H., Montroll, E.W.: Anomalous transit-time dispersion in amorphous solids. Phys. Rev. B **12**(6), 2455–2477 (1975). https://doi.org/10.1103/PhysRevB.12.2455. http://link.aps.org/doi/10.1103/PhysRevB.12.2455

Schirotzek, A., et al.: Observation of Fermi polarons in a tunable Fermi liquid of ultracold atoms. Phys. Rev. Lett. **102**(23), 230402 (2009). https://doi.org/10.1103/PhysRevLett.102.230402

Schleich, W.P.: Quantum optics in phase space. Wiley-VCH, Berlin (2001). https://books.google.
es/books?id=2jUjQPW-WXAC&printsec=frontcover&hl=it&source=gbs_ge_summary_r&
cad=0#v=onepage&q&f=false. ISBN 978-3527294350

Schlosshauer, M.A.: Decoherence and the quantum-to-classical transition. Springer, The Frontiers
Collection (2007). http://www.springer.com/physics/quantum+physics/book/978-3-540-35773-
5. ISBN 9783540357759

Schlosshauer, M.: Decoherence, the measurement problem, and interpretations of quantum mechanics. Rev. Mod. Phys. **76**(4), 1267–1305 (2005). https://doi.org/10.1103/RevModPhys.76.1267.
http://link.aps.org/doi/10.1103/RevModPhys.76.1267

Schmidt, R., et al.: Fermi polarons in two dimensions. Phys. Rev. A **85**(2), 021602 (2012). https://
doi.org/10.1103/PhysRevA.85.021602. https://link.aps.org/doi/10.1103/PhysRevA.85.021602

Shashi, A., et al.: Radio-frequency spectroscopy of polarons in ultracold Bose gases. Phys. Rev. A
89(5), 053617 (2014). https://doi.org/10.1103/PhysRevA.89.053617. http://link.aps.org/doi/10.
1103/PhysRevA.89.053617

Shchadilova, Y.E., et al.: Polaronic mass renormalization of impurities in Bose-Einstein condensates: correlated Gaussian-wave function approach. Phys. Rev. A **93**(4), 043606 (2016a). https://
doi.org/10.1103/PhysRevA.93.043606. https://link.aps.org/doi/10.1103/PhysRevA.93.043606

Shchadilova, Y.E., et al.: Quantum dynamics of ultracold Bose polarons. Phys. Rev. Lett. **117**(11),
113002 (2016b). https://doi.org/10.1103/PhysRevLett.117.113002. https://link.aps.org/doi/10.
1103/PhysRevLett.117.113002

Smoluchowski, M.V.: Drei Vorträge über Diffusion, Brownsche Bewegung und Koagulation von
Kolloidteilchen. Zeitschrift für Physik **17**, 557–585 (1916)

Smoluchowski von, M.: Kinetische Theorie der Brownsche Bewegung und der Suspensionen. Ann.
Phys. **21**(756) (1906)

Spethmann, N., et al.: Dynamics of single neutral impurity atoms immersed in an ultracold gas. Phys.
Rev. Lett. **109**(23), 235301 (2012). https://doi.org/10.1103/PhysRevLett.109.235301. https://
link.aps.org/doi/10.1103/PhysRevLett.109.235301

Subaşi, Y., et al.: Equilibrium states of open quantum systems in the strong coupling regime. Phys.
Rev. E **86**(6), 061132 (2012). https://doi.org/10.1103/PhysRevE.86.061132. http://link.aps.org/
doi/10.1103/PhysRevE.86.061132

Săndulescu, A., Scutaru, H.: Open quantum systems and the damping of collective modes in deep
inelastic collisions. Ann. Phys. **173**(2), 277–317 (1987)

Svedberg, T.: Über die Eigenbewegung der Teilchen in Kolloidalen Lösungen. Von The Svedberg.
Zeitschrift für Elektrochemie und angewandte physikalische Chemie **12**(47), 853–860 (1906).
ISSN: 0005-9021; https://doi.org/10.1002/bbpc.19060124702. http://dx.doi.org/10.1002/bbpc.
19060124702

Tolić-Nørrelykke, I.M., et al.: Anomalous diffusion in living yeast cells. Phys. Rev. Lett.
93(7), 078102 (2004). https://doi.org/10.1103/PhysRevLett.93.078102. http://link.aps.org/doi/
10.1103/PhysRevLett.93.078102

Tuziemski, W.H.J., Korbicz, J.K.: Objectivisation in simplified quantum brownian motion models. Photonics **228–240**, (2015). https://doi.org/10.1016/0370-1573(88)90023-3. http://www.
sciencedirect.com/science/article/pii/0370157388900233

Ullersma, P.: An exactly solvable model for Brownian motion: I Derivation of the Langevin equation.
Physica **32**, 27 (1966). https://doi.org/10.1016/0031-8914(66)90102-9

Vacchini, B.: Completely positive quantum dissipation. Phys. Rev. Lett. **84**(7), 1374–1377 (2000).
https://doi.org/10.1103/PhysRevLett.84.1374. http://link.aps.org/doi/10.1103/PhysRevLett.84.
1374

de Vega, I., Alonso, D.: Dynamics of non-Markovian open quantum systems. Rev. Mod. Phys.
89(1), 015001 (2017). https://doi.org/10.1103/RevModPhys.89.015001. https://link.aps.org/doi/
10.1103/RevModPhys.89.015001

Volosniev, A.G., Hammer, H.-W., Zinner, N.T.: Real-time dynamics of an impurity in an ideal Bose
gas in a trap. Phys. Rev. A **92**(2), 023623 (2015). https://doi.org/10.1103/PhysRevA.92.023623.
https://link.aps.org/doi/10.1103/PhysRevA.92.023623

Volpe, G., et al.: Influence of noise on force measurements. Phys. Rev. Lett. **104**, 170602 (2010)

von Waldenfels, W.: A Measure Theoretical Approach to Quantum Stochastic Processes. Lecture Notes in Physics. Springer, Heidelberg (2014). ISBN 9783642450815

Weedbrook, C., et al.: Gaussian quantum information. Rev. Mod. Phys. **84**(2), 621–669 (2012). https://doi.org/10.1103/RevModPhys.84.621. http://link.aps.org/doi/10.1103/RevModPhys.84.621

Weigel, A.V., et al.: Ergodic and nonergodic processes coexist in the plasma membrane as observed by single-molecule tracking. Proc. Nat. Acad. Sci. **108**(16), 6438–6443 (2011). https://doi.org/10.1073/pnas.1016325108. http://www.pnas.org/content/early/2011/03/28/1016325108.abstract

Weiss, U.: Quantum Dissipative Systems. World Scientific, Singapore (2008)

Wiseman, H.M., Munro, W.J.: Comment on dissipative quantum dynamics with a Lindblad functional. Phys. Rev. Lett. **80**(25), 5702–5702 (1998). https://doi.org/10.1103/PhysRevLett.80.5702. http://link.aps.org/doi/10.1103/PhysRevLett.80.5702

Wodkiewicz, K., Eberly, J.H.: Markovian and non-Markovian behavior in two-level atom fluorescenc. Ann. Phys. **101**(2), 574–593 (1976). https://doi.org/10.1016/0003-4916(76)90023-3. http://www.sciencedirect.com/science/article/pii/0003491676900233

Zakowicz, W., Rzazewski, K.: Collective radiation by harmonic oscillators. J. Phys. A: Math., Nucl. Gen. **7**(7), 869 (1974). http://stacks.iop.org/0301-0015/7/i=7/a=012

Zurek, W.H.: Quantum Darwinism. Nat. Phys. **3**, 181–188 (2009). https://doi.org/10.1016/0370-1573(88)90023-3. http://www.sciencedirect.com/science/article/pii/0370157388900233

Zurek, W.H.: Decoherence, einselection, and the quantum origins of the classical. Rev. Mod. Phys. **75**(3), 715–775 (2003). https://doi.org/10.1103/RevModPhys.75.715. http://link.aps.org/doi/10.1103/RevModPhys.75.715

Zwanzig, R.: Memory effects in irreversible thermodynamics. Phys. Rev. **124**(4), 983–992 (1961). https://doi.org/10.1103/PhysRev.24.983. http://link.aps.org/doi/10.1103/PhysRev.124.983

Printed in the United States
By Bookmasters